Interpreting Quantitative Data

D0163746

Melissa J. Faulkner

Interpreting Quantitative Data with SPSS

Rachad Antonius

SAGE Publications
London • Thousand Oaks • New Delhi

© Rachad Antonius 2003

First published 2003

Reprinted 2004

Apart from any fair dealing for the purposes of research or
private study, or criticism or review, as permitted under
the Copyright, Designs and Patents Act, 1988, this publication
may be reproduced, stored or transmitted in any form, or by
any means, only with the prior permission in writing of the
publishers, or in the case of reprographic reproduction, in
accordance with the terms of licences issued by the Copyright
Licensing Agency. Inquiries concerning reproduction outside
those terms should be sent to the publishers.

SAGE Publications Ltd
1 Oliver's Yard
55 City Road
London EC1Y 1SP

SAGE Publications Inc
2455 Teller Road
Thousand Oaks, California 91320

SAGE Publications India Pvt Ltd
B-42 Panchsheel Enclave
Post Box 4109
New Delhi 100 017

British Library Cataloguing in Publication data

A catalogue record for this book is available from
the British Library

ISBN 0 7619 7398 2
ISBN 0 7619 7399 0 (pbk)

Library of Congress Control Number: 2002 102 782

Typeset by C&M Digital (P) Ltd., Chennai, India
Printed in Great Britain by TJ International Ltd, Padstow, Cornwall

CONTENTS

ACKNOWLEDGMENTS

A number of contingencies have presided over the coming about of this book ... and a number of people have played a role in its birth. I would like to acknowledge their input here.

First of all, there was Louise Corriveau, who first encouraged me to work on a manual of quantitative methodology for the social sciences. She introduced me to a statistician, the late Robert Trudel, and together we wrote *Méthodes quantitatives appliquées aux sciences humaines*. Our long and elaborated discussions on every aspect of the book, both from the point of view of statistics and from the point of view of pedagogy, helped shape my views of how that subject matter ought to be taught. I should point out the role played then by Mr. Charles Dufresne, whose advice on the form, content and organization of that book was an extremely formative experience.

As computers were becoming available for our classes, we started using software packages to teach this course. After experimenting with several packages, the administration and instructors of the course opted for SPSS. We had manuals to teach SPSS, but we did not have a manual to teach quantitative methods for the social sciences with the help of a package such as SPSS.

I thus wrote a series of class notes in quantitative methodology using SPSS. The focus was on methodology, and SPSS was a tool, not an end in itself. With the comments of my students and my colleagues at Champlain College – St-Lambert, these notes gradually evolved into an experimental manuscript. This book is the result of these efforts.

I would like to thank my colleagues – both in the Mathematics department and in the Methodology module – for letting me teach the course, and for testing the manual in their classes. The administration at Champlain was quite supportive, and accommodated my needs in terms of teaching and of granting (unpaid ...) leaves of absence when that was required. They also provided numerous signs of encouragement and appreciation of this work.

As any author knows, writing a book puts a lot of stress on the daily organization of family life. I would like to thank my spouse, Y.G., and my children, Marc and Gabriel, for putting up with my constant preoccupation with statistics and methodology during moments when I should have been available to interact with them.

Finally, I would like to thank all the staff at SAGE and Keyword Publishing Services for their professional editing job.

Rachad Antonius
Montreal, May 20, 2002

Thanks are extended to SPSS Inc. for permission to use copies of SPSS for Windows screens.

SPSS is a registered trademark of SPSS Inc.

For information about SPSS contact:

SPSS Inc., 233 S. Wacker Drive, 11th Floor,
Chicago, IL 60606, USA
Tel: +1 (312) 651 3000
www.spss.com

Statistics Canada information is used with the permission of the Minister of Industry, as Minister responsible for Statistics Canada. Information on the availability of the wide range of data from Statistics Canada can be obtained from Statistics Canada's Regional Offices, its World Wide Web site at http://www.statcan.ca, and its toll-free access number 1-800-263-1136.

Every effort has been made to trace the copyright holders for material used in this book, but if any have been inadvertently overlooked the Publishers will be pleased to make the necessary arrangements at the first opportunity.

FOREWORD TO THE STUDENT

Objectives of the Book

This book corresponds to an introductory course in quantitative methods for the social or human sciences. It explains the basic statistical procedures used in social research, and places their use in the general research process of which they are a part. The theoretical explanations are accompanied by applied exercises that use the SPSS software. After studying the theoretical material and doing the SPSS exercises, students should be able to collect their own data, enter it in SPSS, analyze it, interpret the results, and present such results in summary reports.

What follows consists of more details on the contents of the book, how it is structured, and how to make the best use of it.

The book deals with the production, presentation, analysis, and interpretation of quantitative data, all presented and conceptualized as part of a social research process. Thus the interpretation of statistical results takes as much importance in the book as the explanation of the formulas used to compute such results. The book makes use of elementary statistical techniques that are explained and put into use with the help of the Statistical Package for Social Sciences software, known as SPSS. No prior knowledge of SPSS or of statistics is assumed, and this statistical package is explained in 14 lab sessions that guide the student through the elementary functions of the program, which correspond to the theoretical material shown in the first part of the book.

We will examine the research process as a whole, and see where and how quantitative methods are used. Students who study this book are expected to acquire the practical abilities needed to produce data files, to organize them, to carry out statistical computations with them, to present their results, and to interpret them correctly. In addition to these abilities, students are also expected to acquire some of the theoretical knowledge that will allow them to use quantitative methods in an appropriate manner, and to understand their power and, more importantly, their limits.

More specifically, the book explains:

- what is a quantitative data file;
- how to read it, that is, how to interpret its immediate meanings;
- how to collect data for social research;
- how to organize data into data files;

- how to analyze data;
- how to interpret the results of the analysis;
- how to present the results and their interpretations.

Some of these questions have straightforward answers. Others will require a detailed examination of how quantitative methods fit in a research process, which generally involves aspects that are not quantitative. Students should keep in mind that it is the *qualitative* aspects of a research that are the crucial elements guaranteeing the rigor of the research and the reliability of the results. Indeed, it is very easy, with the availability of powerful software, to carry out complex computations that will produce numerical results. But such results could be meaningless, or even completely erroneous, if they do not rest on solid theoretical foundations, which are usually expressed in qualitative terms. To give an example, consider the notion of intelligence. At the beginning of the twentieth century, some scientists measured with great accuracy the shapes and sizes of people's skulls, and counted the bumps on the skulls of the individuals, noting their size and location. They believed that these measures gave the unmistakable signs of intelligence. But they were wrong. And no statistical treatment of their data, however complex and sophisticated, could compensate for the theoretical flaws that were at the basis of such data.

Upon completing the course, students should have acquired the abilities listed at the beginning of every chapter. A brief version of this list is given below. It could be used by the students as a checklist, to keep track of the main objectives they feel they have achieved. These abilities are grouped into five broad categories: explaining the steps of a social research, understanding quantitative methods and their place in social research, producing an electronic data file, analyzing a data file using SPSS, and producing a descriptive report. Here are the details.

Checklist of Abilities to be Acquired

EXPLAINING THE STEPS OF A SOCIAL RESEARCH
Students should be able to explain the steps of a social research. They should know:

- the broad steps needed to complete any research (quantitative or qualitative);
- the role and importance of theory in orienting an empirical research;
- the main research designs used to produce quantitative data;
- how to select a random sample or a systematic sample;
- the basic steps to be taken in conducting a survey;
- the structure of the basic experimental design in social research;
- the ethical guidelines that should be followed in research on human subjects.

UNDERSTANDING QUANTITATIVE METHODS AND THEIR PLACE IN SOCIAL RESEARCH
Students should know:

- the basic vocabulary of statistics and quantitative methods;
- the type of variables and of measurement scales;

- how concepts are operationalized with the help of indicators;
- the main types of quantitative research (surveys, experiments, and archival research);
- the basic definition of descriptive and inferential statistics;
- the different uses of the term 'statistics';
- what an electronic data file looks like, and how to identify cases and variables;
- how to use printed and electronic databases; how to read a report.

PRODUCING AN ELECTRONIC DATA FILE
Students should be able to:

- determine how to organize the variables and how to determine their types;
- specify their characteristics and define them in SPSS;
- enter the data;
- save and print a data file and an output file.

ANALYZING A DATA FILE USING SPSS
Students should be able to:

- present data (frequency tables; charts), describe the shape of a distribution (symmetry; skewness) and produce these outputs with SPSS;
- determine which measures and charts are appropriate, depending on the measurement level of the variable;
- produce and interpret the various descriptive measures;
- explain the differences in the significance and the uses of the mean and the median;
- produce and read a frequency table;
- use the table of areas under a normal curve;
- analyze statistical association with different procedures, depending on the measurement level of the variables;
- produce and read a two-way table (manually and with SPSS);
- produce and interpret a coefficient of correlation and a scatter plot;
- produce and interpret confidence statements (such as the results of a poll);
- reproduce the logical reasoning underlying hypothesis testing;
- perform and interpret simple t-tests.

PRODUCING A DESCRIPTIVE REPORT
Students should be able to:

- write a report using a word-processing software;
- copy tables and charts into a word processor;
- explain, in plain English, the meanings of the numerical results produced with SPSS;
- write and present the report of a quantitative analysis: describing the data file, describing the variables and the scales used, presenting and describing the data, presenting elementary interpretations (the distributions; statistical association, inference, etc.).

Study Tips

Mastering this material requires several learning activities:

- reading the material;
- doing the exercises;
- doing the SPSS labs;
- reviewing the material and integrating the various components of the acquired knowledge.

Here are some suggestions for each of these four learning activities.

READING THE MATERIAL

The material in this book is written in a rather concise form. It is very important to read it more than once, and with attention. The first reading allows you to understand the scope of a chapter and its principal aim. Some of the fine points may be missed during that first reading. A second reading allows you to consolidate what you have learned in the first reading and to capture some of the details missed during the first reading. It is a good idea to read with a pencil and paper at hand, and to write down important definitions, or formulas, or some idea that seems to hold the key to understanding subsequent ideas, so as to remember these elements more easily. Finally, a third reading is recommended towards the end of the course, after you have covered most chapters and gone through the SPSS lab sessions. You will see then that you read into the text elements that you had not seen in the first readings.

DOING THE EXERCISES

The exercises at the end of each chapter are important to help you really understand the material explained in the chapter. Some exercises are reading exercises. They force you to look for some specific details in the text and to be able either to mention them as is, or to reformulate them in your own words. Some are computational exercises. In order to solve these problems, you must have understood the procedure used to perform a certain computation, and you must be able to reproduce it in a specific situation. While doing the exercises, you may have to go to the main text to find a definition or an explanation of the procedure.

DOING THE SPSS LABS

The SPSS labs are an integral part of learning the material in this book, as the data files you will be working with contain a wealth of concrete examples that illustrate further the theoretical material explained in the first part of the book.

The labs are all structured in a similar way: some procedures are explained in detail and their results shown. You should perform these procedures yourself on your computer as you follow the explanations, making sure that what you get on your screen corresponds to what is explained in the book. This should give you an understanding of how the procedure works. You are then asked to answer a question that requires using the same procedure for a different variable or maybe a different data file altogether. After going through all the labs, you should be able to perform an elementary statistical analysis of a data file on your own.

REVIEWING THE MATERIAL

This is a very important step. Learning does not progress linearly, but rather in a spiral movement. After seeing a concept, you need to see other concepts to which it is applied, or that follow it. You then come back to the first concept and understand it at a deeper level.

Four tools are provided to review the material. The first is given by the set of keywords at the end of every chapter. You can review a chapter by trying to give the definition of each keyword mentioned at its end. The labs are also a way of reviewing the material, as you are required to use concepts introduced in the first part of the book. A *synthesis* given at the end of the theoretical part of the book constitutes a transversal review of the book, whereas the various tasks required to write a report about a data file cut across several chapters. This synthesis will also help to integrate the material learned, that is, to recapitulate it in a context that is different from the one where it was learned, and to combine several techniques learned in various chapters in a single portion of the analysis. The review questions given at the end of the first part of the book, before the labs, constitute the fourth tool for reviewing the material covered in the book.

Finally, it is hoped that the approach used here will show the relevance of statistical techniques to make meaningful comments about both society and individuals. The book is intended as an introductory first course in quantitative methods. Most students in social or human sciences will be expected to pursue the learning of quantitative methods through subsequent, more advanced courses. We hope that this book will motivate them in this endeavor and will make their task both more efficient and more pleasant.

FOREWORD TO THE INSTRUCTOR

Some specific comments need to be added to the Foreword, concerning the material included in the book, the references, and the pedagogical approach used.

The Material Included

This book is designed for an introductory course in quantitative methods applied to the social and human sciences. It makes a synthesis between statistics, research methodology, and the use of SPSS, and these three dimensions are integrated and combined to get the students to learn how to write a satisfactory statistical report.

All the *statistical* material included is quite basic. It is restricted to the elementary notions of descriptive statistics and inferential statistics. No mathematical proofs are given, but the meanings of the formulas and the concrete ways of interpreting and using them are explained with some degree of detail. SPSS is taught here not as an end in itself, and the SPSS labs should not be thought of as reference material for SPSS. Rather, the SPSS labs were written in such a way as to allow the student to perform the calculations explained in the first part of the book, and to produce and interpret some of the basic outputs. My experience in the testing stages of this manual is that students acquire enough familiarity with SPSS to be able to figure out on their own how to perform procedures that were not taught in this course. It should be pointed out that we did not use the SPSS syntax at all. We briefly mentioned what it is and how to paste it automatically, but all the procedures shown are menu-driven and are explained by the point-and-click method.

For the sake of completeness, a chapter and a lab have been included on accessing online databases. These constitute a wealth of resources for the social scientist, and they include not only aggregate data on a large spectrum of relevant topics, but also data sets containing individual data that can be retrieved and analyzed.

The material included is oriented towards getting the student to be able to write a simple analysis of the data contained in a data file, including basic descriptive measures, some graphical illustrations, simple estimations, and simple hypothesis tests. The SPSS procedures explained in the labs have been determined in accordance with that aim. A whole chapter has been included, right after the chapter on descriptive statistics, to discuss the way a descriptive report should be written, and a synthesis at the end of the theoretical part expands that chapter to include inferential procedures.

There has been a systematic attempt to situate the use of statistics within a comprehensive view of the research process, which means that the conditions under which a statistical method is used, its limitations, and the interpretation of the results it produces are always discussed as the method is presented. However, the material in this book is focused on quantitative methods, and any reference to the research process aims simply at situating the methods shown in their social science context. Ideally, students would be taking another course on research methodology, preferably after completing this course. In the testing stages of this manual, however, some students had taken the general methodology course either before this one or concurrently with it, and in both cases the results were satisfactory, as the students were applying the methods learned here almost concurrently as they were learning them in the general research methods course.

Although the book does not discuss the philosophy of social science at all, we adopt an implicit anti-positivist orientation. It does not imply a rejection of statistical methods, but rather an understanding of their limitations, and an understanding of the fact that whatever is measured is not a faithful representation of 'reality', but rather a representation of our reconstruction of what we perceive in reality. We may hope that such representations are faithful, but that remains in the realm of wishes, not of science. What follows from this attitude is an increased care in expliciting the assumptions that lie beneath any conceptual construct, and a readiness to question claims about the validity of such constructs.

In terms of time use, we suggest distributing the SPSS exercises evenly throughout the semester. While the SPSS exercises make more sense when linked to the theoretical material, they can also be covered almost independently. Instructors will note that the SPSS exercise that consists in creating a data file is not given at the beginning, but is listed as lab 9. This is because the importance of the correct definition of variables can be understood only after seeing that the correct statistical procedures to be used depend on the characteristics of the variable, and these characteristics must be factored in the very definition of the variable in the SPSS data file. Detailed comments on time use will be found on the website devoted to the pedagogical aspects of this book (see below).

References

The statistical material included in this book is basic, and it has been part of the folklore of statistics and quantitative methods for a long time. Therefore, it did not make sense to give references for the basic formulas that were mentioned, such as that of the standard deviation or of the margin of error when estimating a parameter. Instead, we have included in every chapter a section called *Suggestions for further reading* that could orient the reader to more advanced textbooks, or to textbooks focused on a specific discipline.

We have also included a list of references after the Synthesis chapter, at the end of the book. The references included have been selected with the following criteria in mind. Every reference corresponds to one criterion or more.

1. They went further in the subject, covering topics that are not covered in this book, or covering the same topics in more depth and detail.
2. They were basic supporting references for the statistical formulas.
3. They covered specific disciplinary approaches.
4. They were complementary, covering qualitative research for instance.
5. They included interesting critical views of statistics in research.
6. They constituted reference material for SPSS.

These references were not included in the suggested reading provided with the various chapters.

Additional Material

Some additional material has been posted on the website (http://www.champlain college.qc.ca/antonius). This includes a detailed discussion of the pedagogical approach used in this book, some additional exercises, some specific pedagogical tips concerning timing and presentation, and solutions to the exercises and labs.

The Pedagogical Approach

What follows is an underpinning of five pedagogical principles on which this approach is based. We have tried to apply them in a comprehensive way now made possible by the dissemination of computer technology in the classroom. The method presented here may achieve its full pedagogical potential when students have access on a weekly basis to a computer lab equipped with the SPSS program, or have acquired the academic version of SPSS. It is also quite useful to have, at least occasionally, access to a multimedia projector in the classroom, linked to an SPSS-equipped computer. Under these conditions, every theoretical discussion can be illustrated by a direct display of SPSS data or SPSS output on a large screen in the classroom. Hundreds of variables become instantly accessible to the instructor, and the class can move to an interactive mode, allowing the instructor to produce instantly charts or tables in response to a question from a student. This supposes of course a certain familiarity both with SPSS and with the data files used as examples, but the new versions of SPSS make it so user-friendly that this familiarity is quickly acquired. Moreover, an instructor in a specific discipline can supply data files from his or her own research and fit them in the pedagogical process proposed in this manual. This instant accessibility to many examples in the classroom allowed us to write a rather concise text, focusing on the logic of the statistical approach and on its abstract structure, knowing that numerous examples can be accessed directly, in class, and in an interactive manner.

Here then are the basic principles on which our approach was based. This approach is partly the result of long experience of teaching two subjects: mathematics to student audiences that are not inclined towards it, and sociology of foreign societies whose inner workings and logic are not readily accessible to North-American student audiences. In both cases, the questions of induction vs. deduction, of relevance, and of intuitive knowledge and intuitive understanding pose themselves in an acute way. We found that this experience, and the lessons learned from it, became quite useful when teaching quantitative methods to students not inclined towards statistics.

Principle 1:
Knowledge is constructed by a mixture of inductive and deductive modes of thinking, which have different roles in the learning process.

Inductive modes of thinking are more effective when the subject is new, or when its relevance has not yet been established, or when the level of maturity needed for learning it deductively has not been achieved. Induction corresponds to the stage of discovery, of changes of paradigms, and of understanding of a new subject.
Deductive modes of thinking are more effective at the stage of establishing a logical order between concepts, ideas, and theories, and of sorting out potentially brilliant – but false – intuitions from correct facts, that is, at the stage of establishing *proofs*. This is also the stage of effective and efficient organization of knowledge.

We assert that for this course, the inductive modes of reasoning should take precedence (chronologically) over the deductive modes.

Some of the difficulty encountered by students learning quantitative methods and statistics is precisely the fact that certain ways of reasoning are presented in a deductive fashion, making their logical underpinnings very solid, but meaning nothing to a majority of the students. A related principle is the question of relevance explained below.

Principle 2:
The establishment of the **relevance** of a statement must **precede** the demonstration of its **truth**.

This is another way of saying that the answer to a question becomes meaningful only after the question itself has been understood, assimilated, integrated, and incorporated into one's view of the subject.

To show the **relevance** of a question means to show:

- its meaning and significance;
- its importance in comparison to other questions;
- the consequences that different answers to the question might have on the problem we are dealing with.

We claim that for most books on statistics, the material covered is organized according to a deductive logic, which draws its internal consistency from the fact that it is the logic of proof, even when proofs of theorems are not given. For the purpose of training future statisticians this is absolutely crucial, of course. But for training users of statistics in the social sciences (and maybe in some other domains as well) we claim that another order of presentation is required. Such a logical order would involve first an elaboration of the **meaning** of what is to be demonstrated, then of its **relevance**, then of the **practical ways of applying it**, and only last, if time permits, the **proof** of the truth of a statement or of the exactness of a method. At that point, the importance of the proof must be asserted, together with the fact that *the only thing that justifies presenting the method in this way is that a proof of its exactness exists somewhere else*, and that statisticians have actually proven that it works. This does not imply, though, that the proof itself must be presented.

To illustrate our point, let us take the example of confidence statements. We can help students understand, as a first step, the very notion of a confidence statement in a way that does not involve any exact calculation of margin of error or of probability of error. The calculations are introduced only after the concept itself has been understood.

For instance, it can be explained in class that if a representative sample (a notion to be used intuitively at this point) of students in a college is chosen, and if their height was measured and its average found to be 170 cm, we could guess that the average height for the whole population is expected to be *around* 170 cm, maybe somewhere between say 165 cm and 175 cm unless, by an extraordinary strike of bad luck, all the students we picked at random turned out to include the whole basketball team of the college. We can then explain the concept of margin of error without being able to compute such a margin yet, and we can explain that we need to find a way to determine whether the margin of error should be from 165 to 175 cm, or between any other values. Methods for calculating the margin of error should be introduced only after the very concept is understood, and then they can refer to the notion of sampling distribution, which is the basis for calculating the margin of error associated with some probability level. Then, we could mention that this method of calculation can be proven mathematically to be correct. A mathematician may object by saying that the very notion of the margin of error can be understood correctly only when you show the proof of why it works (which gives you at the same time the method for computing it). We claim that this is true when a certain level of maturity has been achieved. We claim that our method contributes to helping the students achieve such a level of maturity. A subsequent, more advanced course in statistics for social science students could certainly include more sophisticated mathematical ideas.

Principle 3:
Intuition could be developed through such a course, and not supposed to be already developed.

This goal can be achieved by two methods: the frequent use of **diagrams** and the constant reformulation of conclusions in plain English, that is, the translation of numerical results into full English sentences and vice versa.

Regarding the importance of diagrams, a systematic effort has been made to include diagrams that establish relationships between concepts. Thus students do not learn individual concepts, but networks of concepts that structure the field of quantitative methods. Establishing such networks of concepts has the double function of making them more relevant and of playing the role of a mnemonic device. Hence, many diagrams have been introduced in the book.

The translation of numerical results into full sentences in plain English is an essential aspect of the exercises students must perform to really understand the material. Thus, exercises that consist in reading a two-way table (or cross-tabulation) are essential. Students should be able to determine whether a sentence like 'the men of this sample are more likely than the women to behave in a manner X' follows from a two-way table or not.

Principle 4:
The order of succession of the steps of a research project is not necessarily the same as the order in which students can learn them.

For instance, creating a data file comes logically before the analysis of data files, but we assert that many of the methodological issues that need to be addressed when creating a data file can be understood only *after* you have had an opportunity to work with a data file and perform some of the statistical analyses. This is particularly true of some of the issues that relate to the codebook and to the operational definition of a concept. Thus, the creation of a data file in SPSS is shown only after the section on descriptive statistics.

As a result, the method used here is based on the following succession: *Intuitive notions of statistics are presented first, and they are stretched to their maximum logical limits*. Only when the notions are well understood intuitively can one see the logical limits of such notions. *Formalization and rigorous definitions are then introduced as a response to the limits of intuitive notions*. They come at the end of the discussion of a subject, not at its beginning.

Principle 5:
Intellectual maturity should not be considered a prerequisite for that course (although when present, it would allow a much faster progression). It should be consciously built and developed in the mind of students through the learning activities of this course.

Intellectual maturity is the capacity to understand the *relevance* of an argument, and *to find connections* between different arguments and different ideas. It is the capacity to situate the argument as a whole in a larger conceptual entity that confers relevance to it and gives it its significance. This significance allows us to understand the logical necessity of the various parts of the argument and its logical coherence. This kind of contextual information is usually not included in the argument itself. This means that maturity is achieved when things are understood without having to be said in full detail. *Maturity is also the capacity of filling in details that may be missing when they are not logically necessary from a deductive point of view but they are necessary from the point of view of the relevance of the argument.* Thus a mathematical proof may be presented in a comprehensive way, and be perfectly understandable to a mature mathematician, but may appear at the same time totally incomprehensible to a beginner.

Maturity is acquired by experience and it is founded on analogies between a situation at hand and other situations seen previously and understood, that may present some degree of similarity. The lessons drawn from such analogies are often not explicit, but they require some kind of integration of previous knowledge: an *understanding* of their contextual meaning, then a *retention* of such a meaning followed by a *transfer* to a different context. Thus, at its lowest levels, maturity is specific to a field of knowledge, but it may become transferable to other fields of knowledge.

The implication of the above remarks for this book is that maturity has to be consciously built and developed in the mind of students by a systematic discussion of all the elements that confer relevance to the question studied, and to the logical elements that contribute to establishing the answers. Developing maturity in quantitative methods may be facilitated by weekly access to a computer lab. Numerous statistical examples can be seen in a very short period of time, and if properly directed, students can see at the click of a mouse the connections between some features of a distribution and their statistical properties.

The principles outlined above have immediate consequences on the way the material is organized. In the first chapter, for instance, we start describing an SPSS data file, as it first appears when the program is opened. We learn how to read it and interpret the information in it, with an approach that may appear to be positivist at first. But immediately after, we raise questions about how the data file was produced in the first place, and show the relevance of many questions that will gradually be answered as the course unfolds. We end up questioning positivist approaches by showing that there are many ways of defining a concept and of presenting it, depending on how the concept is conceptualized in the theoretical framework that is used.

Finally, we hope that these comments constitute a contribution to a dialogue about pedagogy among colleagues, in which the author is eager to engage. Any comments are welcome, and the author would make it a point to respond to any discussion initiated on these points.

Rachad Antonius
Champlain College – Saint-Lambert
and
The University of Montreal
Montreal, Canada
antoniur@magellan.umontreal.ca

1

THE BASIC LANGUAGE OF STATISTICS

This chapter is an introduction to statistics and to quantitative methods. It explains the basic language used in statistics, the notion of a data file, the distinction between descriptive and inferential statistics, and the basic concepts of statistics and quantitative methods.

After studying this chapter, the student should know:

- the basic vocabulary of statistics and of quantitative methods;
- what an electronic data file looks like, and how to identify cases and variables;
- the different uses of the term 'statistics';
- the basic definition of descriptive and inferential statistics;
- the type of variables and of measurement scales;
- how concepts are operationalized with the help of indicators.

Introduction: Social Sciences and Quantitative Methods

Social sciences aim at studying social and human phenomena as rigorously as possible. This involves describing some aspect of the social reality, analyzing it to see whether logical links can be established between its various parts, and, whenever possible, predicting future outcomes.

The general objective of such studies is to understand the patterns of individual or collective behavior, the constraints that affect it, the causes and explanations that can help us understand our societies and ourselves better and predict the consequences of certain situations. Such studies are never entirely objective, as they are inevitably based on certain assumptions and beliefs that cannot be demonstrated. Our perceptions of social phenomena are themselves subjective to a large extent, as they depend on the *meanings* we attribute to what we observe. Thus, we *interpret* social and human phenomena much more than we describe them, but we try to make that interpretation as objective as possible.

Some of the phenomena we observe can be *quantified*, which means that we can translate into numbers some aspects of our observations. For instance, we can quantify population change: we can count how many babies are born every year in a given country, how many people die, and how many people migrate in or out of the country. Such figures allow us to estimate the present size of the population, and maybe even to predict how this size is going to change in the near future. We can quantify psychological phenomena such as the degree of stress or the rapidity of response to a stimulus; demographic phenomena such as population sizes or sex ratios (the ratio of men to women); geographic phenomena such as the average amount of rain over a year or over a month; economic phenomena such as the unemployment rate; we can also quantify social phenomena such as the changing patterns of marriage or of unions, and so on.

When a social or human phenomenon is quantified in an appropriate way, we can ground our analysis of it on figures, or statistics. This allows us to describe the phenomenon with some accuracy, to establish whether there are links between some of the variables, and even to predict the evolution of the phenomenon. If the observations have been conducted on a sample (that is, a group of people smaller than the whole population), we may even be able to generalize to the whole population what we have found on a sample.

When we observe a social or human phenomenon in a systematic, scientific way, the information we gather about it is referred to as *data*. In other words, **data** is information that is collected in a systematic way, and organized and recorded in such a way that it can be interpreted correctly. Data is not collected haphazardly, but in response to some questions that the researchers would like to answer. Sometimes, we collect information (that is, data) about a character or a quality, such as the mother tongue of a person. Sometimes, the data is something measurable with numbers, such as a person's age. In both cases, we can treat this data numerically: for instance we can count how many people speak a certain language, or we can find the average age of a group of people. The procedures and techniques used to analyze data numerically are called *quantitative methods*. In other words, **quantitative methods** are procedures and techniques used to analyze data numerically; they include a study of the valid methods used for collecting data in the first place, as well as a discussion of the limits of validity of any given procedure (that is, an understanding of the situations when a given procedure yields valid results), and of the ways the results are to be interpreted.

This book constitutes an introduction to quantitative methods for the social sciences. The first chapter covers the basic vocabulary of quantitative methods. This vocabulary should be mastered by the student if the remainder of the book is to be understood properly.

Data Files

The first object of analysis in quantitative methods is a **data file**, that is, a set of pieces of information written down in a codified way. Figure 1.1 illustrates what an **electronic data file** looks like when we open it with the SPSS program.

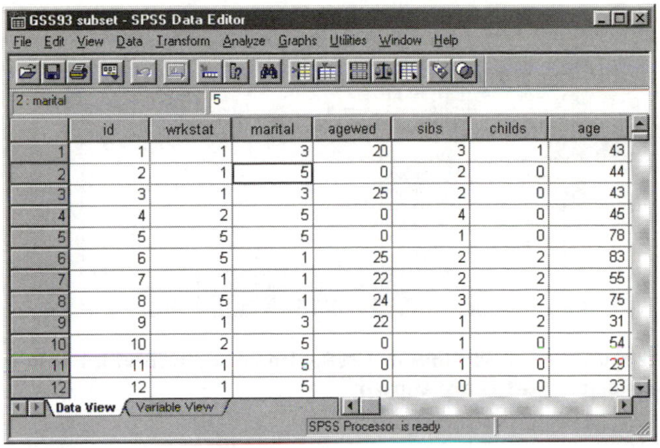

Figure 1.1 **The Data window in SPSS version 10.1. © SPSS. Reprinted with permission.**

This data file was created by the statistical software package SPSS Version 10.1, which will be used in this course. The first lab in the second part of this manual will introduce you to SPSS, which stands for *Statistical Package for the Social Sciences*. On the top of the window, you can read the name of the data file: **GSS93 subset**. This stands for **Subset of the General Social Survey**, a survey conducted in the USA in 1993.

When we open an SPSS data file, two views can be displayed: the Data View or the Variable View. Both views are part of the same file, and one can switch from one view to the other by clicking on the tab at the bottom left of the window.

The Data View: The information in this data view is organized in rows and columns. Each row refers to a **case**, that is, all the information pertaining to one individual. Each column refers to a **variable**, that is, a character or quality that was measured in this survey. For instance, the second column is a variable called **wrkstat**, and the third is a variable called **marital**.

But what are the meanings of all these numbers and words? A data file must be accompanied by information that allows a reader to interpret (that is, understand) the meanings of the various elements in it. This information constitutes the **codebook**. In SPSS, we can find the information of the codebook by clicking the word **Variables...** under the **Utilities** menu. We get a window listing all the variables contained in this data file. By clicking once on a variable, we see the information pertaining to this variable:

- the short name that stands on the top of the column;
- what the name stands for (the **label** of the variable);
- the numerical type of the variable (that is, how many digits are used, and whether it includes decimals);
- other technical information to be explained later;
- and the **Value Labels**, that is, what each number appearing in the data sheet stands for.

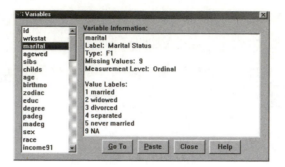

Figure 1.2 **The Variables window in SPSS. The codes and value labels of the variable Marital Status are shown**

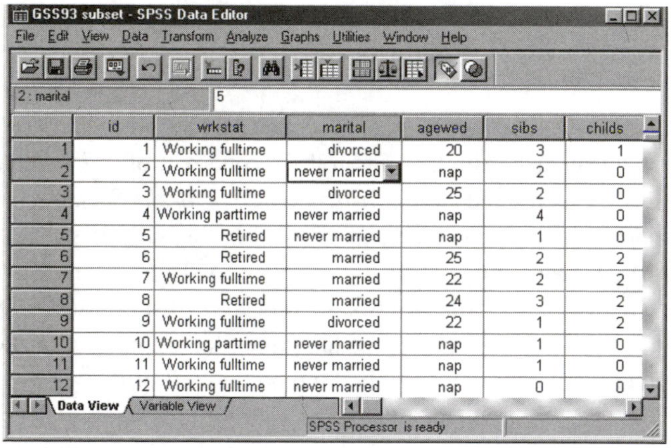

Figure 1.3 **The Data View window in SPSS when the Show Labels command is ticked in the View menu. The value labels are displayed rather than the codes**

Figure 1.2 shows the codes used for the variable Marital Status.

You may have noticed that:

> 1   stands for   married
> 2   stands for   widowed
> 3   stands for   divorced
> etc.

The numbers 1, 2, 3, etc. are the **codes,** and the terms married, widowed, divorced, etc. are the **value labels** that correspond to the various codes. The name *marital*, which appears at the top of the column, is the **variable name**. *Marital Status* is the **variable label**: it is a usually longer, detailed name for the variable. When we print tables or graphs, it is the variable labels and the value labels that are printed.

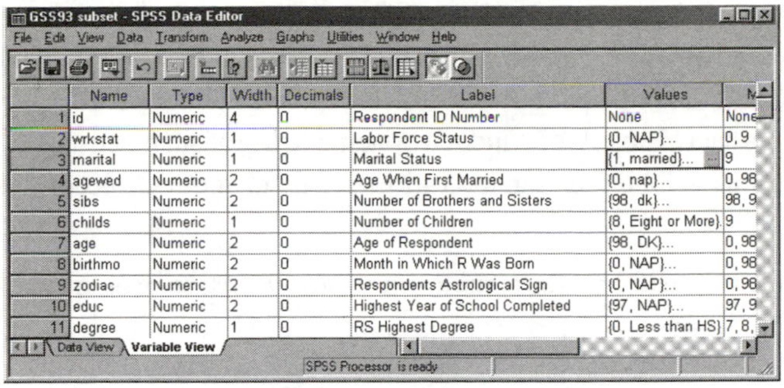

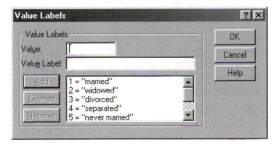

Figure 1.4 **The Variable View window in SPSS. The variables are listed in the rows, and their properties are displayed**

Figure 1.5 **The Value Labels window in SPSS. In this window it is possible to add new codes and their corresponding value labels, or to modify or delete existing ones**

There is a way of showing the value labels instead of the codes. This is done by clicking **Value Labels** under the **View** menu. The Data View window looks now as shown in Figure 1.3.

We can see that case number 4, for example, is a person who works part time, and who has never been married. To understand the precise meaning of the numbers written in the other cells, we should first read the variable information found in the codebook for each of the variables.

In version 10.0 and version 11 of SPSS, you can read the information pertaining to the variables in the Variable View. By clicking on the tab for Variable View, you get the window shown in Figure 1.4.

In the Variable View, no data is shown. You can see, however, all the information pertaining to the variables themselves, each variable being represented by a line. The various variable names are listed in the first column, and each is followed by information about the corresponding variable: the way it is measured and recorded, its full name, the values and their codes, etc. All these terms will be explained in detail later on. The label, that is, the long name of the variable **marital**, is **Marital Status**. By clicking on the **Values** cell for the variable **marital**, the window shown in Figure 1.5 pops up.

We can see again the meanings of the codes used to designate the various marital statuses. We can now raise a number of questions: How did we come up with this data? What are the rules for obtaining reliable data that can be interpreted easily? How can we analyze this data? Table 1.1 includes a systematic list of such questions. The answers to these questions will be found in the various chapters and sections of this manual.

Table 1.1 **Some questions that arise when we want to use quantitative methods**

Questions	Chapters
How did we come up with this data? What are the questions we are trying to answer? What is the place of quantitative analysis in social research, and how does it link up with the qualitative questions we may want to ask? What is the scientific way of defining concepts and operationalizing them?	1. The Basic Language of Statistics
How do we conduct social research in a scientific way? What procedures should we follow to ensure that results are scientific? What are the basic types of research designs? How do we go about collecting the data?	2. The Research Process
Once collected, the data must be organized and described. How do we do that? When we summarize the data what are the characteristics that we focus on? What kind of information is lost? What are the most common types of shapes and distributions we encounter?	3. Univariate Descriptive Statistics
	5. Normal Distributions
What are the procedures for selecting a sample? Are some of them better than others?	6. Sampling Designs
Some institutions collect and publish a lot of social data. Where can we find it? How do we use it?	7. Statistical Databases
Sometimes we notice coincidences in the data: for instance, those who have a higher income tend to behave differently on some social variables than those who do not. Is there a way of describing such relationships between variables, and drawing their significance?	8. Statistical Association
Sometimes the data comes from a sample, that is, a part of the population, and not the whole population. Can we generalize our conclusions to the whole population on the basis of the data collected on a sample? How can this be done? Is it precise? What are the risks that our conclusions are wrong?	9. Statistical Inference: Estimation 10. Statistical Inference: Hypothesis Testing

The Discipline of Statistics

The term *statistics* is used in two different meanings: it can refer to the *discipline* of statistics, or it can refer to the *actual data* that has been collected.

As a scientific discipline, the object of *statistics* is the numerical treatment of data that pertain to a large quantity of individuals or a large quantity of objects. It includes a general, theoretical aspect which is very mathematical, but it can also include the study of the concrete problems that are raised when we apply the theoretical methods to specific disciplines. The term *quantitative methods* is used to refer to methods and techniques of statistics which are applied to concrete problems. Thus, the difference between statistics and quantitative methods is that the latter include practical concerns such as finding solutions to the problems arising from the collection of real data, and interpreting the numerical results as they relate to concrete situations. For instance, proving that the mean (or average) of a set of values has certain mathematical properties is part of statistics. Deciding that the mean is an appropriate measure to use in a given situation is part of quantitative methods. But the line between statistics and quantitative methods is fuzzy, and the two terms are sometimes used interchangeably. In practice, the term statistics is often used to mean quantitative methods, and we will use it in that way too.

The term *statistics* has also a different meaning, and it is used to refer to the actual data that has been obtained by statistical methods. Thus, we will say for instance that the latest statistics published by the Ministry of Labor indicate a decrease in unemployment. In that last sentence, the word *statistics* was used to refer to data published by the Ministry.

Populations, Samples, and Units

Three basic terms must be defined to explain the subject matter of the discipline of statistics:

- unit (or element, or case),
- population, and
- sample.

A **unit** (sometimes called **element**, or **case**) is the smallest object of study. If we are conducting a study on individuals, a unit is an individual. If our study were about the health system (we may want to know, for instance, whether certain hospitals are more efficient than others), a unit for such a study would be a hospital, not a person.

A **population** is the collection of all units that we wish to consider. If our study is about the hospitals in Quebec, the population will consist of all hospitals in Quebec. Sometimes, the term *universe* is used to refer to the set of all individuals under consideration, but we will not use it in this manual.

Most of the time, we cannot afford to study each and every unit in a population, due to the impossibility of doing so or to considerations of time and cost. In this case, we study a smaller group of units, called a **sample**. Thus, a sample is *any* subset (or subgroup) of our population.

STATISTICS	
Descriptive statistics	**Inferential statistics**
It aims at describing a situation by summarizing information in a way that highlights the important numerical features of the data. Some of the information is lost as a result. A good summary captures the essential aspects of the data and the most relevant ones.	It aims at inferring (that is, drawing conclusions about) some numerical character of a population when only a sample is given. The inference always implies a margin of error and a probability of error. Inferences based on representative samples have a higher chance of being correct. A random sample is more likely to be representative.
MEASURES OF CENTRAL TENDENCY They answer the question: What are the values that represent the bulk of the data in the best way? Mean, median, mode.	**ESTIMATION** It is based on the distinction between sample and population. It consists in guessing the value of a measure on a population (i.e. a **parameter**) when only the value on the sample is known (the **statistic**). Opinion polls are always based on estimations: the survey is conducted on a representative sample, and its results are generalized to the population with a margin of error and a probability of error.
MEASURES OF DISPERSION They answer the question: How spread out is the data? Is it mostly concentrated around the center, or spread out over a large range? Standard deviation, variance, range.	
MEASURES OF POSITION They answer the question: How is one individual entry positioned with respect to all the others? Percentiles, deciles, quartiles.	**HYPOTHESIS TESTING** It is also based on the distinction between sample and population, but the process is reversed: We make a hypothesis about a population parameter. On that basis, we predict a range of values a variable is likely to take on a representative sample. Then we go and measure the sample. If the observed value falls within the predicted range, we conclude that the hypothesis is reasonable. If the observed value falls outside the predicted range, we reject our hypothesis.
MEASURES OF ASSOCIATION They answer the question: If we know the score of an individual on one variable, to what extent can we successfully predict how he is likely to score on the other variable? Correlation coefficient (r)	

Figure 1.6   The discipline of statistics and its two branches, descriptive statistics and inferential statistics

The distinction between sample and population is absolutely fundamental. Whenever you are doing a computation, or making any statement, it must be clear in your mind whether you are talking about a sample (a group of units generally smaller than the population) or about the whole population.

The discipline of statistics includes two main branches:

- descriptive statistics, and
- inferential statistics.

The following paragraphs explain what each branch is about. Refer also to Figure 1.6. Some of the terms used in the diagram may not be clear for now, but they will be explained as we progress.

Descriptive Statistics

The methods and techniques of descriptive statistics aim at summarizing large quantities of data by a few numbers, in a way that highlights the most important numerical features of the data. For instance, if you say that your average GPA (grade point average) in secondary schooling is 3.62, you are giving only one number that gives a pretty good idea of your performance during all your secondary schooling. If you also say that the *standard deviation* (this term will be explained later on) of your grades is 0.02, you are saying that your marks are very consistent across the various courses. A standard deviation of 0.1 would indicate a variability that is 5 times bigger, as we will learn later on. You do not need to give the detailed list of your marks in every exam of every course: the average GPA is a sufficient measure in many circumstances. However, the average can sometimes be misleading. When is the average misleading? Can we complement it by other measures that would help us have a better idea of the features of the data we are summarizing? Such questions are part of descriptive statistics.

Descriptive statistics include measures of central tendency, measures of dispersion, measures of position, and measures of association. They also include a description of the general shape of the distribution of the data. These terms will be explained in the corresponding chapters.

Inferential Statistics

Inferential statistics aim at generalizing a measure taken on a small number of cases that have been observed, to a larger set of cases that have not been observed. Using the terms explained above, we could reformulate this aim, and say that inferential statistics aim at generalizing observations made on a sample to a whole population. For instance, when pre-election polls are conducted, only one or two thousand individuals are questioned, and on the basis of their answers, the polling agency draws conclusions about the voting intentions of the whole population. Such conclusions are not very precise, and there is always a risk that they are completely wrong. More importantly, the sample used to draw such conclusions must be a *representative sample*, that is, a sample in which all the relevant qualities of the population are adequately represented. How can we ensure that a sample is representative? Well, we can't. We can only increase our chances of selecting a representative sample if we select it randomly. We will devote a chapter to sampling methods.

Inferential statistics include estimation and hypothesis testing, two techniques that will be studied in Chapters 9 and 10.

A few more terms must be defined to be able to go further in our study. We need to talk a little about variables and their types.

Variables and Measurement

A variable is a characteristic or quality that is observed, measured, and recorded in a data file (generally, in a single column). If you need to keep track of the country of birth of the individuals in your population, you will include in your study a variable called *Country of birth*. You may also want to keep track of the nationality of the individuals: you will then have another variable called *Nationality*. The two variables are distinct, since some people may carry the nationality of a country other than the one they were born in. Here are some examples of variables used widely in social sciences:

Socio-demographic variables
Age
Sex
Religion
Level of education
Highest degree obtained
Marital status
Country of birth
Nationality
Mother tongue

Psychological variables
Level of anxiety
Stimulus response time
Score obtained in a personality test
Score obtained in an aptitude test

Economic variables
Working status
Income
Value of individual assets
Average number of hours of work
  per week

Variables that refer to units other than the individual
Number of hospitals in a country
Percentage of people who can read
Percentage of people who completed
  high school
Total population
Birth rate
Fertility rate
Number of teachers per 1000 people
Number of doctors per 10,000
  people
Population growth
Predominant religion

You may have noticed that some of these variables refer to qualities (such as mother tongue) and others refer to quantities, such as the total population of a country. In fact, we can distinguish two basic **types of variables**:

1. **quantitative variables;**
2. **qualitative variables**.

Quantitative variables are characteristics or features that are best expressed by numerical values, such as the age of a person, the number of people in a household, the size of a building, or the annual sales of a product. Qualitative variables are characteristics or qualities that are not numerical, such as mother tongue, or country of origin. The scores of the individuals of a population on the various variables are called the **values** of that variable.

Example

Suppose you have the information shown in Table 1.2 about five students in your college.

Table 1.2 **Examples of qualitative and quantitative variables**

Name	Age	Program of Study	Grade Point Average
John	19	Social Science	3.78
Mary	17	Pure and Applied Science	3.89
Peter	18	Commerce	3.67
Colette	19	Office Systems Technology	3.90
Suzie	20	Graphic Design	3.82

There are three variables: **Age** (quantitative), **Program of Study** (qualitative), and **Grade Point Average** (quantitative).

The values, or scores, taken by the individuals for the variable **Age** are 17, 18, 19 (twice), and 20. The values taken for the variable **Program of Study** are Social Science, Pure and Applied Science, Commerce, Office Systems Technology, and Graphic Design. Qualitative variables are sometimes referred to as **categorical** variables because they consist of categories in which the population can be classified. For instance, we can classify all students in a college into categories according to the program of study they are in.

Careful attention must be given to the way observations pertaining to a variable are *recorded*. We must find a system for recording the data that is very clear, and that can be interpreted without any ambiguity. Consider, for instance, the following characteristics: age, rank in the family, and mother tongue. The first characteristic is a quantity; the second is a rank, and the third is a quality. The systems used to record our observations about these characteristics will be organized into three **levels of measurement**:

- measurement at the **nominal** level;
- measurement at the **ordinal** level; and
- measurement at the **numerical scale** level.

Each level of measurement allows us to perform certain statistical operations, and not others.

The **nominal level of measurement** is used to measure qualitative variables. It is the simplest system for writing down our observations: when we want to measure a characteristic at the nominal level, we establish a number of categories in such a way that each observation falls into one and only one of these categories. For example, if you want to write down your observations about mother tongue in the Canadian context, you may have the following categories:

- English,
- French,
- Native, and
- Other.

Depending on the subject of your research, you may have more categories to include other languages, or you may want to make a provision for those who have two mother tongues.

It is important to note that when a variable is measured at the nominal level, the categories must be

- exhaustive, and
- mutually exclusive.

The categories are said to be **exhaustive** when they include the whole range of possible observations, that is, they exhaust all the possibilities. That means that every one of the observations can fit in one of the available categories. The categories are said to be **mutually exclusive** if they are not overlapping: every observation fits in only one category. These two properties ensure that the system used to write down the observations is clear and complete, and that there are no ambiguities when recording the observations or when reading the data file. Table 1.3 displays examples of measurements made at the nominal level.

Qualitative variables must be measured at the nominal level.

The **ordinal level of measurement** is used when the observations are organized in categories that are *ranked*, or *ordered*. We can say that one category precedes another, but we cannot say by how much exactly (or if we can, we do not keep that information). Here too the categories must be exhaustive and mutually exclusive, but in addition you must be able to compare any two categories, and say which one precedes the other (or is bigger, or better, etc.). Table 1.4 displays examples of variables measured at the ordinal level.

Table 1.3 **Examples of variables measured at the nominal level**

Variable	Categories used
Sex	Male
	Female
Place of birth	The country where the survey is
	counducted
	Abroad
Work status	Working full-time
	Working part-time
	Temporarily out of work
	Unemployed
	Retired
	Housekeeper
	Other

Table 1.4 **Examples of variables measured at the ordinal level**

Variable	Ranked Categories
Rating of a restaurant	Excellent
	Very good
	Acceptable
	Poor
	Very poor
Rank among siblings	First child
	Second child
	etc.
Income	High
	Medium
	Low

The scale used to write down an ordinal variable is often referred to as a **Likert scale**. It usually has a limited number of ranked categories: anywhere from three to seven categories, sometimes more. For instance, if people are asked to rate a service as:

- ❑ Excellent
- ❑ Very good
- ❑ Good
- ❑ Poor
- ❑ Very poor,

the proposed answers constitute a five-level Likert scale.

Another example of a Likert scale, this time with four levels, is provided by the situations where a statement is given, and respondents are asked to say whether they:

❑ Totally agree
❑ Agree
❑ Disagree
❑ Totally disagree.

A variable measured at the ordinal level could be either qualitative or quantitative. In Table 1.4, the variable **Income** is quantitative, and the variable **Rating of a Restaurant** is qualitative, but they are both measured at the ordinal level. For a variable measured at the ordinal level, we can say that one value precedes another, but we cannot give an exact numerical value for the difference between them. For instance, if we know that a respondent is the first child and the other is the second child in the same family, we do not keep track of the age difference between them. It could be one year in one case and five years in another case, but the values recorded under this variable do not give us this information: they only give us the rank.

When recording information about categorical variables, the information is usually *coded*. *Coding* is the operation by which we determine the categories that will be recorded, and the codes used to refer to them. For instance, if the variable is **Sex**, and the two possible answers are:

Male
Female,

we usually code this variable as

1   Male
2   Female.

The numbers 1 and 2 are the *codes*, and the categories Male and Female are the *values* of the variable.

When coding a variable, a code must be given to the cases where no answer has been provided by the respondent, or when the respondent refuses to answer (if the answer is judged too personal or confidential, such as the exact income of a person). We refer to these answers as *missing values* and we give them different codes. Lab 9 explains how to handle them in SPSS.

Finally, some variables are measured by a **numerical scale**. Every observation is measured against the scale and assigned a numerical value, which measures a quantity. These variables are said to be **quantitative**. Table 1.5 displays examples of numerical scale variables.

Table 1.5 **Examples of variables measured at the numerical scale level**

Variable	Numerical Scale
Annual income	In dollars, without decimals (no cents)
Age	In years, with no fractions
Age	In years, with one decimal for fractions of a year
Temperature	In degrees Celsius
Time	In years. A starting point must be specified
Annual income	In dollars, to the nearest thousand

Notice that the same variable can be measured by different scales, as shown in the examples above. So, when we use a numerical scale, we must determine the units used (for instance years or months), and the number of decimals used.

Numerical scales are sometimes subdivided into **interval scales** and **ratio scales**, depending on whether there is an absolute zero to the scale or not. Thus, *temperature* and *time* are measured by interval scales, whereas *age* and *number of children* are each measured by a ratio scale. However, this distinction will not be relevant for most of what we are doing in this course, and we will simply use the term **numerical scale** to talk about this level of measurement. The program SPSS that we are going to use simply uses the term **scale** to refer to such variables.

Most statistical software packages include more specific ways of writing down the observations pertaining to a numerical scale. For instance, SPSS will offer the possibility of specifying that the variable is a currency, or a date.

Moreover, it is also possible to group the values of a quantitative variable into **classes**. Thus, when observing the variable *age*, we can write down the exact age of a person in years, or we can simply write the age group the person falls in, as is done in the following example:

- 18 to 30 years
- 31 to 40 years
- 41 to 50 years
- 50 to 60 years
- Over 60.

When we group a variable such as *age* into a small number of categories as we have just done, we must code the categories as we do for categorical variables. For example,

1	would stand for the category 18 to 30 years
2	would stand for the category 31 to 40 years
	etc.

In such situations, we cannot perform the same statistical operations that we do when the values are not grouped. For instance, the mean, or average of the variable *age* is best calculated when the ages are *not* grouped. When we group the values, it is because we want to know the relative importance (that is, the frequency, in percentage) of one group as compared to the others. The information that 50% of the population is under 20 years old in some developing countries is obtained by grouping the ages into *20 years old or less* and *more than 20 years old*. When we collect the data, it is always better to collect it in actual years, since we can easily group it later on in the data file with the help of a statistical software package. In this case, a new column is added to the data file, and it contains the grouped data of the quantitative variable. For example, in the **GSS93 subset** data file that we use in the SPSS labs, you will find two variables for *age*: one is called **age**, and the other one is called **agecat4**. The latter is calculated from the former, by grouping individuals into four age groups. In the column of **agecat4**, the specific age of an individual is not recorded: only the age group of the individual is recorded.

Finally, numerical scales can be either **continuous** or **discrete**. A scale is said to be *continuous* if the observations can theoretically take any value over a certain range, including fractions of a unit. For instance, age, weight, length are continuous variables because they are not limited to specific values, and they can take any value within a certain range. A variable is said to be *discrete* if it can take only a limited number of possible values, but not values in between. For instance, the variable *Number of children* is measured by a discrete scale because it can only be equal to a whole number: 0, 1, 2, etc.

Importance of the Level of Measurement

The level of measurement used for a variable depends on whether it is qualitative or quantitative.

Qualitative variables must be measured at the nominal or ordinal level. They cannot be measured at the numerical scale level, even when their categories are coded with numbers. For instance, as shown above, we usually code the variable *Sex* as follows:

1   Male
2   Female.

In this case, **the numbers 1 and 2 have no numerical value**. They are simply codes. It is shorter to write 1 than *male*, and we could have assigned the numbers differently. If you ask SPSS to compute the mean (or average) for a variable coded in this way, you *will* get a numerical answer. But you must always keep in mind that such a numerical answer is *totally meaningless* because the level of measurement of that variable is nominal. The numbers used to record the information are simply codes.

Quantitative variables are usually measured by a numerical scale, but they could be measured at the ordinal level also. For instance, if you have the annual income of an individual, you may treat it as a numerical scale, but you could also group the values into Low, Medium and High income and treat the variable at the ordinal level.

When you perform a statistical analysis of data, it is very important to pay attention to the level of measurement of each variable. Some statistical computations are appropriate only to a given level of measurement, and should not be performed if the variable is measured at a different level.

Concepts, Dimensions, and Indicators

We often want to observe social phenomena that are too abstract and complex to be expressed by a single variable. Suppose for instance that we want to observe and measure the degree of *religious inclination* (or the tendency of a person towards religion) in a given social group. Religious inclination can be manifested in many ways: people may have or not have certain *beliefs* about their religion; they may also perform or not certain *rituals* such as attending religious services, fasting, praying, etc.; they may also *seek the advice of the religious leadership* on important decisions, or ignore such leadership; finally, they may seek to look at everything from the point of view of religion, and *apply the teachings* of their religion in their daily lives, or ignore them. All these aspects are not found all the time in all individuals. Some individuals may have strong beliefs, while avoiding the religious services. Other may attend all services while being skeptical about some of the religious dogma. The way to handle this complexity is to subdivide the concept of *religious inclination* into dimensions, which are themselves measured by several indicators. If we were to study religious inclination in the Catholic religion, we would get a set of dimensions and indicators that would look as in Table 1.6 (we are simplifying the issues a little, of course).

The items listed on the right-hand side of Table 1.6 are **indicators** of the concept of *religious inclination*. None of them, taken alone, is a measure of religious inclination, but each of them constitutes *one* aspect of it. Indicators that are seen as similar are grouped together to form one *dimension* of the concept. And finally the various dimensions, taken together, capture the concept as a whole. This way of breaking down a complex concept into dimensions and indicators is called the **operationalization** of the concept. As an illustration, we may want to see how economists operationalize the concept of *cost of living*. They estimate the average cost of most of the standard expenses a family of four is expected to incur. The various expenses are divided into main dimensions such as food, housing, transportation, education, and leisure. Each dimension is then subdivided into smaller dimensions; themselves subdivided further until indicators are reached. For instance, food is

Table 1.6 **Example of how a concept can be broken down into dimensions and indicators**

Concept	Dimensions	Indicators
	I. Beliefs	Belief in God
		Belief in the Holy Trinity
		Belief in the main dogma
		etc.
RELIGIOUS INCLINATION	II. Rituals	Attendance of services
		Performing prayers
		Baptizing children
		etc.
	III. Guidance	Consulting the priest about important decisions
		Consulting the official opinions of the church on certain issues such as birth control
		etc.
	IV. Daily life	Being kind and generous to people
		Not cheating others in commercial transactions
		etc.

broken down as: meat, vegetables, milk products, etc., themselves subdivided into specific items such as: tomatoes, lettuce, etc. Finally, for each of these indicators, the increase or decrease in the cost of living is measured against the corresponding cost in some year, called the base year. By combining these indicators, economists are able to measure how the cost of living has changed, on the average, for a family of four.

The way a concept is broken down, or operationalized, into dimensions and indicators depends on the theoretical framework adopted for a study. Researchers may not agree on how to operationalize a concept, and you will find in the literature different studies that operationalize concepts in completely different ways, because they rely on different theoretical frameworks.

Summary

Quantitative methods are procedures and techniques for collecting, organizing, describing, analyzing, and interpreting data. In this chapter we have learned the basic vocabulary used to talk about quantitative methods. Data is organized into electronic data files with the help of statistical packages. A data file contains the values taken by a number of cases (which are the units of the population under study) over some variables. Every row represents a case, while every column represents a variable. The units in the data file usually form a sample, which is itself a subset of the whole population. Sometimes, the data file refers to the whole population.

The variables can be either qualitative or quantitative. The system used to record the information is called a measurement scale. There are three levels of measurement: nominal, ordinal and numerical (interval or ratio). The level of measurement of a variable will determine what statistical procedures can be performed, and what kind of graphs must be used to illustrate the data. When a concept is complex, it is not measured directly. It is usually broken down into dimensions and indicators, which are then combined to provide a single measure.

The statistical procedures themselves fall into two broad categories: descriptive statistics and inferential statistics. Descriptive statistical techniques aim at describing the data by summarizing it, while inferential statistical techniques aim at generalizing to a whole population what has been observed on a sample.

Keywords

Students should be able to define and explain *all* the following terms.

Data	Case	Continuous numerical scales
Data file	Unit	Discrete numerical scales
Quantitative methods	Sample	Codes
Variable	Population	Coding
Variable label	Level of measurement	Codebook
Value	Nominal level	Statistics (the two meanings)
Value label	Ordinal level	Descriptive statistics
Variable type	Numerical level (interval or ratio)	Inferential statistics
Quantitative variable	Exhaustive categories	Dimensions of a concept
Qualitative variable	Mutually exclusive categories	Indicators of a concept
	Likert scale	Operationalization of a concept

Suggestions for Further Reading

Blalock Jr., Hubert M. (1982) *Conceptualization and Measurement in the Social Sciences*. London; Sage Publications.

Norusis, Marija J. (1998) *SPSS 8.0 Guide to Data Analysis*. Upper Saddle River, NJ: Prentice Hall.

Rosenbaum, Sonia (1979) *Quantitative Methods and Statistics: A Guide to Social Research*. Beverly Hills: Sage Publications.

Trudel, Robert and Antonius, Rachad (1991) *Méthodes quantitatives appliquées aux sciences humaines*. Montréal: CEC.

EXERCISES

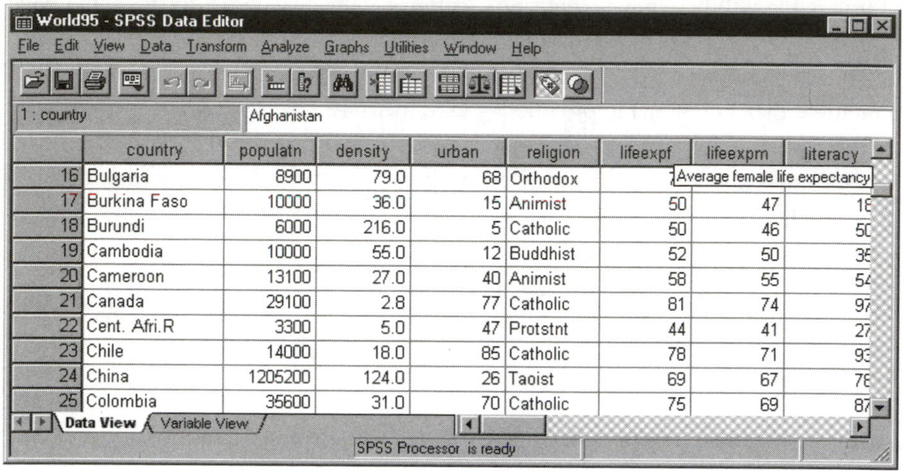

	country	populatn	density	urban	religion	lifeexpf	lifeexpm	literacy
16	Bulgaria	8900	79.0	68	Orthodox	Average female life expectancy		
17	Burkina Faso	10000	36.0	15	Animist	50	47	18
18	Burundi	6000	216.0	5	Catholic	50	46	50
19	Cambodia	10000	55.0	12	Buddhist	52	50	35
20	Cameroon	13100	27.0	40	Animist	58	55	54
21	Canada	29100	2.8	77	Catholic	81	74	97
22	Cent. Afri.R	3300	5.0	47	Protstnt	44	41	27
23	Chile	14000	18.0	85	Catholic	78	71	93
24	China	1205200	124.0	26	Taoist	69	67	78
25	Colombia	35600	31.0	70	Catholic	75	69	87

World95 - SPSS Data Editor — File Edit View Data Transform Analyze Graphs Utilities Window Help — 1 : country — Afghanistan — Data View / Variable View — SPSS Processor is ready

Figure 1.7

1.1   Consider the window shown in Figure 1.7, obtained when you open a data file with SPSS. The following questions can be answered by inspecting this window carefully.

   (a)   What is the name of the file?
   (b)   What are the names of the variables shown?
   (c)   The variable label is given for one of the variables. What is it?
   (d)   What are the value labels that you could see for the variable **religion**?
   (e)   For each variable shown, say whether it is qualitative or quantitative, and indicate the correct level of measurement to be used.
   (f)   List all the values shown for the variable *population* (the actual name used in this data file is *populatn*), indicating the unit of measurement that is used.
   (g)   Describe case number 20 on the basis of what is shown.

1.2   The following three questions are asked in a questionnaire, and the answers are recorded at the nominal level. Indicate whether the categories offered are exhaustive and mutually exclusive. Indicate whether these categories are adequate for the society you live in, and if they are not, propose another set of categories.

   1.   Marital status: Married
                           Divorced
                           Single

2. Language spoken at home: English
                                French
                                Other

3. What method of transportation do you mostly use to come to school? (Choose only one):

    Public transportation
    Private car (alone or shared)
    Bicycle
    Walking
    Other

1.3 Determine whether the following variables are qualitative or quantitative.

age	height	marital status
program of study	country of origin	nationality
number of children	income	ownership of home (Yes/No)
value of your house	religion you were raised in	GPA in high school

1.4 Determine the level of measurement used in the following cases.

(a) Annual Income: $ _ _ _ _ _ _ . _ _

(b) Annual Income: 1. Low (less than $20,000)
                       2. Medium (more than $20,000 but less than $50,000)
                       3. High (more than $50,000)

(c) Sex: 1. Male
                       2. Female

(d) Language(s) spoken: 1. French
                            2. English
                            3. German
                            4. Other

(e) Number of languages spoken: 1. One language
                                 2. Two languages
                                 3. Three languages
                                 4. More than three

2

THE RESEARCH PROCESS

This chapter explains the basic steps involved in a research process. It also explains briefly three specific types of research that produce quantitative data: surveys, experiments, and archival research.

After studying this chapter, the student should know:

- the broad steps needed to complete any research (quantitative or qualitative);
- the role and importance of theory in orienting an empirical research;
- the main types of quantitative research (surveys, experiments, and archival research);
- the basic steps to be taken in conducting a survey;
- the structure of the basic experimental design in social research;
- the ethical guidelines that should be followed in research on human subjects.

The basic object of study of this book is the quantitative treatment of data. In order to interpret the data correctly, however, and to draw the correct conclusions from the analysis we must know how this data was produced in the first place, and in response to which questions. The method used to produce the data and the theoretical framework of the research are important elements that allow us to interpret the data correctly. In other words, to make sense of quantitative data as a tool for understanding social issues, we must have a more global view of the research process and of the place of quantitative data in that process, which is essentially qualitative.

The present chapter examines the wider process of research and how the production of quantitative data fits in it. If conducted properly, this process will also allow a correct interpretation of the data produced. We will therefore examine what constitutes an acceptable process for conducting social research. There are many ways of organizing and presenting the steps of social research. But they all boil down to the same fundamental steps. We will summarize them in the following way.

Main Steps in Social Research

1. Determine the general research question.
2. Conduct a thorough literature review.
3. Determine the theoretical framework and the specific research question.
4. Determine a research design.
5. Collect the data.
6. Analyze the data.
7. Interpret the results.
8. Draw general conclusions and raise further questions.

We will now examine each of these steps in some detail.

1. The General Research Question

This is the starting point. In order to start a research project, you must first determine a general theme and a general research question. The general research question must be chosen so that it is relevant either from some theoretical perspective or from the point of view of its concrete applications. You should also do some preliminary reading and investigations to find out what are the relevant questions or issues that are related to that theme. This should help you elaborate a **problematique**, that is, a whole set of related questions, together with a discussion of their relevance and of their meaning. After you have done such a preliminary investigation, you may be ready to formulate or reformulate the general research question to take into account what you found in the preliminary readings. The general research question is sometimes formulated as a hypothesis to be proved or disproved: it is then called the **research hypothesis**.

> ### Example
>
> For example, the general research theme may be: *Students and work*. Here are some general questions that are related to that theme.
>
> - What is the *situation* now for college students: Do most of them work? For how many hours per week on the average? What kinds of jobs do they tend to take? Do they work mostly during the summer when they do not have courses, or do they also work during the school year? etc. These types of questions lead to a descriptive study, which aims at representing the situation of students' work as observed.
> - What are the *factors* that push students to work (part time or full time)? By factors, we mean: economic, cultural, individual factors, etc., and we also include individual motivations. Do students work out of necessity, to pay for

their basic expenses and tuition fees, or do they work to pay for luxury items such as fashionable clothing and maybe a car? Do people coming from various ethnic backgrounds have a different attitude to student work? Do parents encourage them to be self-sufficient while studying or do they look after their expenses? Is this related to the level of their income or to their culture? Questions of this sort would lead to an explanatory study, not just to a description of what is observed.

• What are the *consequences* of their work on their academic results? We can distinguish here several situations: cases where students work a few hours a week (maybe an afternoon in the weekend) and cases where students work during the week, and for long hours. The consequences themselves may cover a wide range of issues: grades and academic performance, choice of programs, attitudes towards studying, health, social life, etc. This type of question could be either explanatory or predictive, that is, have as an objective to determine the academic results that are expected when we know the working status of a student as well as other background information.

When a researcher starts a research of this kind, it is assumed that she or he is not completely new to the field or to the discipline. Therefore, some familiarity with the issue is assumed, which may have been gained in previous studies or in previous research. Nevertheless, before we decide on the general research question we want to work on, we must conduct a *preliminary literature review*, to see if there are outstanding issues we may have overlooked, and to gain more insights into the subject of research. We could first consult one or two textbooks in the field of education to see how this topic is treated, and what are the important issues that must be considered. We could also consult one of the important journals of education to find a couple of articles on the topic. For instance, we may find in the literature a discussion of *gender* as an important factor in the attitudes of people to study and work, in the sense that the motivations as well as the patterns and the consequences of gainful employment among female students differ significantly from those among male students.

Having conducted that quick investigation, combined with previous familiarity with the field, we are in a position to ask a general question. That question will constitute a guide for the investigation. It will allow us to set for ourselves strategic goals to achieve, without being distracted by the many questions that could arise during the research, some of which may have not been anticipated. For instance, using the example above, the research question could be:

To what extent do cultural values affect people's attitude to gainful employment while studying (as opposed to economic necessity or other factors)?

Having established that general question, we may want to conduct a thorough literature review.

2. The Literature Review

The literature review has the function of allowing us to build on what other researchers have done before us. Knowledge in the social sciences is not the result of one single and isolated mind, no matter how bright it may be. It is the result of collective efforts by scores of people working in the field. We do not want to reinvent the wheel every time we have some insights into a social issue. We should rather take into account what has already been done, and see whether we can add some more by answering new questions, or by extending previous studies to populations not previously studied.

A **literature review** is a systematic reading of the important contributions in the field of study, with a focus on the research question that has been chosen. It aims at:

- establishing the list of authors, papers, books, or scientific reports that address the research question;
- identifying the theoretical approaches that are used and deemed useful in studying the issue;
- establishing the major empirical findings about that particular question and related issues.

Literature reviews are made easy by the new technologies of information. Not only do we have access to computerized catalogs of books and disciplinary journals, but we also have access to searchable lists of articles, with their abstracts or with the full text, on CD-ROMs or directly on the World Wide Web, through the Internet. Moreover, the Internet can be of great help as we can access through it material not available in libraries and resource centers.

3. The Theoretical Framework and the Specific Research Question

After having completed the literature review, we have a global view of the approaches used to study a question, of the authors considered to be important, and of the major findings. We must now specify the main research question and the approach *we* want to adopt to study it. We will have to specify the main concepts that we are going to use and their relationships, as well as the main assumptions that underlie the research. All this constitutes the **theoretical framework**. We could also include a discussion of how we are going to **operationalize** the main concepts, that is, how we are going to define them in concrete terms and how to measure them. In specifying the main research question we also have to specify the population to which it applies, and a time frame as well.

The specific research question may be formulated as a research hypothesis. For example, the general research question mentioned above may evolve into a specific hypothesis such as: '*People coming from traditional cultures tend to work while studying less than people raised in a North-American culture.*'

When all this is done, we can think of the research design, which is the next step.

4. The Research Design

The **research design** is a careful planning of the operations to be done to collect the data in a rigorous, systematic way, in accordance with the methods and ethics of social research. There are several types of research designs used in social research. We will focus on three of them:

1. survey designs,
2. experimental designs, and
3. archival research designs.

Other research designs are more suitable for qualitative data and we will not discuss them now.

SURVEYS

Surveys consist in asking a large number of people some specific questions, or in collecting data about a large number of statistical units[1]. When you conduct a survey, you construct a **questionnaire** (that is, a list of all the questions for which you are seeking answers), you determine the sample or population on which you want to conduct it, you then fill the questionnaires either by direct interviews or by asking people in your sample to fill the questionnaires themselves, or by consulting some appropriate archive. If the survey is about *objects* or *institutions* rather than people (as in a survey about hospitals: each hospital is a statistical unit), one questionnaire is filled for every unit. Surveys can be conducted on samples or on a whole population: in this latter case they are called **censuses**. Most countries have established governmental or para-governmental institutions that compile statistics about the country and conduct general population censuses on a periodic base (5 or 10 years, for example). When a survey is conducted, the following operations must be completed:

- Composition of the questionnaire. This is a very important operation. The questionnaire as a whole must allow you to capture all the information that you are looking for. Each question must be carefully thought of and its formulation must be clear. Many researchers prefer to **test the questionnaire** on a small number of subjects before using it on the sample, to see whether the questions are well understood. The questionnaire is then refined and distributed to the whole sample.
- Composition of the codebook. A codebook contains all the information that allows us to record the answers in a data file and to interpret them.

---

[1] We are using this word in its technical sense: a unit is the most elementary object of study. The collection of all units forms a population. A sample is a part of the population. Refer to the precise definitions of these terms in Chapter 1.

- Determination of the sample. First the *sample size* must be determined, and then the sample must be chosen. If the sample is chosen correctly, that is, if it is large enough, representative, and randomly selected, it can allow us to generalize our results to the whole population. A very important aspect of survey design is therefore the **sampling design**, which is the description of a detailed plan for obtaining a sample. Sampling is a subject of study of its own and we will come back to it in a subsequent chapter.

- Collection of the data. There are many methods for collecting the data. Researchers can interview people directly, or they can ask them to fill the questionnaires themselves. The questionnaires can be hand-delivered, or sent by mail. Every method has its advantages and drawbacks. Two of the important issues involved in data collection are the bias introduced by the data collection method, and the **rate of return** of the questionnaires, that is, the proportion of sent questionnaires that have been filled and returned.

- Data entry, data organization and presentation, and data analysis. Once the data has been collected, it must be entered into a computerized data file. It must also be organized and presented in a way that facilitates its interpretation and its analysis. And then it must be analyzed with the help of a statistical software package, which is the subject of the remainder of this book.

- Interpretation of results. The software package will produce scores of tables and graphs at the click of the mouse. It is very easy to produce outputs that look impressive but could be meaningless. Therefore, it is very important to give adequate attention to the theoretical framework of the research. It is the rigor of the qualitative theoretical framework that allows a correct interpretation of the numerical results of a statistical analysis. If the numerical results are interpreted correctly, we can draw appropriate conclusions from the research.

EXPERIMENTS

Experiments are procedures that allow the observation of people's response to a *treatment* under *controlled circumstances*. When the units of the experiment are people, we refer to them as **subjects** rather than units. Experiments can be conducted on statistical units that are not necessarily people. For instance, you can conduct a statistical experiment on how manufactured objects react to heat or to shocks. The two fundamental ideas behind experiments are the idea of *treatment* and the idea of *control*.

By **treatment**, we mean some experimental condition to which the units are subjected. For instance, you may want to experiment with whether or not a certain medicine has an effect on the subjects. Here the treatment consists in taking that medicine.

Control means that you try to isolate the effect of one specific factor, by making sure that all other factors are constant (or at least very similar) throughout the experiment. If you want to assess accurately the impact of a medicine on the subjects, you must be able to sort out that effect from the effect of other factors. The standard way to do it is to divide your sample into two groups: an experimental group, which takes

Figure 2.1   **A schematic representation of the process of experimentation**

the medicine, and a control group, which goes through the same conditions except for taking the medicine. In order to take into account the psychological effects of taking a medicine, the control group is given a fake medicine, called a **placebo**. A placebo is a neutral substance that looks like the real medicine, except that it has no physical effect. Individuals in the sample do not know whether they are taking the real medicine or the placebo. This way, the *difference* in reaction between the experimental group and the control group can be attributed to the medicine. The basic experimental design can be represented by the diagram of Figure 2.1.

Some variations may be introduced on that basic design. For instance, you could have a **double-blind experiment**, which means that neither the subjects nor the person giving the treatment knows who is getting the real treatment and who is getting the placebo: only the researcher who will analyze the results knows it. This is to minimize the possibility that a subject may guess whether he or she is given the placebo or the real treatment.

Another variation consists in having three or more groups, getting different levels of treatment: one group gets a given dosage, another group may get double that dosage, and the third group may get a placebo.

We could also think of a variation where the groups are not distinct, but they are the same people taking a placebo once and the real thing once on two different runs of the experiment.

The experimental and control groups should not be too different. Sometimes, they are formed of matched pairs: subjects are matched in pairs that are very similar on all the variables that count (for instance, sex, height, weight, language, etc.) and one member of the pair is assigned to the experimental group, and the other to the control group. This kind of experimental design is called a **matched-pair design**.

An **experimental design** is a careful planning of the various aspects of the experiment, including the number of experimental groups to be formed, their composition, the level of treatment needed, etc.

ARCHIVAL RESEARCH

Archival research consists in finding, among the information stored in archives, the data that helps in answering the research question. **Archives** are files that contain data stored previously, sometimes even collected for purposes that may have nothing to do with the research. There are many kinds of archives: the data collected and stored by a governmental statistics agency is an example of one type of archival data. This data has not been collected for a given research project, but rather as a portrait of a given society at a given time. Hundreds of variables can be found: demographic variables, variables concerning employment, data on manufacture and industry, on raw material extraction, on other types of economic activity, on social behavior concerning marriage and divorce, on consumption patterns, etc. A researcher who wants to use this data must first understand what is there, how the variables were defined, how the data is stored, and how to access it. Statistical archives are increasingly accessible through the Internet, but usually, only aggregate data is available through the web (aggregate data is data that has been compiled in tables and charts, and it does not include data about the individual statistical units). Researchers who need more detailed data must either buy it or obtain special permission to access it. We will learn to navigate through the archives of these national agencies later on in this course.

But archives include other kinds of data as well. For instance, universities maintain academic archives through which one can study how the student population has evolved over the years, how the grades and the passing rates have evolved, etc. Public institutions and bodies such as ministries, municipalities, etc. also keep archives.

An archival research design will consist in establishing a list of locations where we can find archives relevant to a given research, then finding out what these archives contain in detail, the quality of the data stored (Is it rigorous? Is it complete? Is it easily accessible? Was it stored in a way that allows us to use it? Etc.) and how to access it. The various archives must be explored to see whether the data is useful, and then a plan to integrate the data from these various archives must be developed.

5. Data Collection

This step consists in actually getting the data, in accordance with the research design determined in the earlier step. The specific ways of collecting data, the specific criteria for determining what is good data (that is, precise and appropriate data), and the specific biases that the researcher must try to avoid all depend on the research design that is used in a given research. Some of these issues will be discussed in subsequent chapters.

6. Data Analysis

Once data is collected, it must be prepared for analysis. This includes entering the data in an electronic file by using some statistical software package. The data must then be summarized, and then analyzed with the help of statistical methods, which is what we will focus on in the rest of this book. Figure 1.6 of Chapter 1, titled Statistics, illustrates the various kinds of operations done in statistical analysis. For this course, the software used to enter and analyze data is SPSS (Version 10.1 or Version 11 for Windows), which stands for *Statistical Package for the Social Sciences*. A Macintosh version of SPSS 10.0 is also available.

Ethical questions

Any research design must be subjected to **ethical guidelines** that ensure that the subjects are not harmed, that they are informed of the purpose and of the conditions of the research and of its potential consequences, that they consent to be part of it, and that their privacy is respected. Every research institution (academic or funding agency) must have a set of ethical guidelines for conducting research under its auspices or on its premises. These guidelines should be taken seriously and followed with care. They can be summarized by four important principles:

- **Informed consent**. This means that the subjects chosen for an experiment or volunteering for it *must be informed* of what they are up to, and of the possible consequences if any. *They must agree* to be part of the research on the basis of this honest and accurate information. It is not ethical to trick subjects into a situation they had not considered or did not want to get into.
- **Confidentiality**. Researchers must take the appropriate means to ensure that information about individuals collected in the course of a research project will remain confidential, and that a reader is not able to trace back any piece of information to a particular individual. Individual names must not be written on questionnaires. They must be given a code known only to the main researcher.
- **Absence of relationship of authority**. Subjects should not be pressured to enroll in a research as a result of a relationship of authority of the researchers over them. This would be a situation of conflict of interest on the part of the researchers, and that situation should be avoided. It must be made clear to every subject that participation in a research project is absolutely voluntary.
- **Freedom to withdraw at any time**. Subjects enrolling in a research project must be able to withdraw at any time without any pressure exerted on them to stay, and they must be informed by the researcher that this is their right.

SPSS produces tables, charts, and numerical statistical measures at the click of the mouse. In order to be useful, these numerical results must be interpreted, that is, their meanings must be specified. This course will include learning about the interpretation of the basic statistical measures used in the social sciences.

7. Interpretation of Results

The numerical results obtained with the help of SPSS, together with the statistical interpretation of their meanings, must then be linked to the qualitative research question. We are not talking here of *statistical* interpretation (that was part of data analysis) but of the *qualitative* interpretation of the numerical results. The meaning and relevance of the numerical results must be specified at that point, and conclusions must be drawn about the general question that was discussed.

8. General Conclusions and Further Questions

An empirical research may have repercussions beyond the direct findings that were established through it. First, these findings can either confirm or disprove the theories that were used in the theoretical framework. Secondly, society is a whole, and social phenomena are interrelated. Thus findings in one given sector of human activities raise new questions, either as a follow-up to the research question, or as new questions that apply to other fields of investigation. For instance, new findings about how gender affects learning styles may trigger research about how gender differences are socially constructed. In other words, a good way to conclude a research is to open up new perspectives that affect future research in the same field as well as in other areas of research. You may find in your readings that some research paper has been labeled 'seminal', meaning that it has inspired many researchers and that it has engendered many research papers which relied on it as a starting point.

Summary

We have drawn a broad and general picture of the research process. A deeper understanding of its various aspects takes a long time, and will be possible only after you have participated in one or more research projects. The general picture outlined above should, however, serve as a grid, or as a general framework, that will help you structure the information and understanding you will acquire in the future. The steps explained above are summarized in Figure 2.2 at the end of this chapter.

Keywords

General research question	Literature review	Problematique
Theoretical framework	Research design	Survey designs
Experimental design	Archival research design	Census
Rate of return	Subjects	Treatment
Control group	Placebo	Double-blind experiment
Matched-pair design	Informed consent	

Suggestions for Further Reading

Babbie, Earl R. (1998) *The Practice of Social Research*, 8th edn. Belmont: Wadsworth.
Bailey, Kenneth (1994) *Methods of Social Research*, 4th edn. New York: Free Press.
Bechhofer, Frank (2000) *Principles of Research Design in the Social Sciences*. New York: Routledge.
Converse, Jean M. and Presser, Stanley (1986) *Survey Questions: Handcrafting the Standardized Questionnaire*. Beverly Hills, CA: Sage Publications.

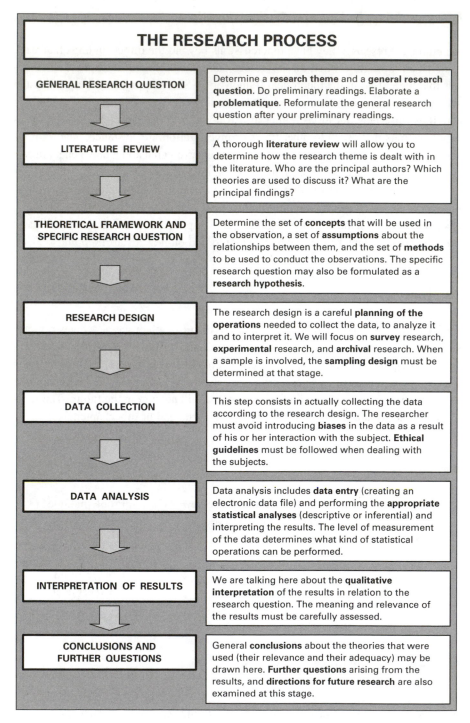

Figure 2.2   **A schematic representation of the research process**

Creswell, John W. (1994) *Research Design: Qualitative & Quantitative Approaches*. Thousand Oaks, CA: Sage.

De Vaus, David A. (1991) *Surveys in Social Research*, Third Edition. London: UCL Press.

Gliner, Jeffrey A. and Morgan, George A. (2000) *Research Methods in Applied Settings: An Integrated Approach to Design and Analysis*. Mahwah, NJ: Lawrence Erlbaum.

Hedrick, Terry E., Bickman, Leonard and Rog, Debra J. (1993) *Applied Research Design: A Practical Guide*. Newbury Park, CA: Sage.

Miller, Delbert (1991) *Handbook of Research Design and Social Measurement*. Newbury Park, CA: Sage.

EXERCISES

2.1   List the main steps of a research process and explain them briefly (2 to 3 lines each).

2.2   Explain in some detail each of the main steps of a research process.

2.3   Explain the steps of an experiment where you want to test the effect of playing music on the performance of students in an exam. Construct an experimental design with one control group and four experimental groups: two groups are made to listen to soft music (low volume and high volume) and two groups to rock music (low volume and high volume). Draw a diagram to illustrate the experiment, similar to Figure 2.1.

3

UNIVARIATE DESCRIPTIVE STATISTICS

This chapter explains how data concerning one variable can be summarized and described, with tables and with simple charts and diagrams.

After studying this chapter, the student should know:

- the basic types of univariate descriptive measures;
- how the level of measurement determines the descriptive measures to be used;
- how to interpret these descriptive measures;
- how to read a frequency table;
- the differences in the significance and the uses of the mean and the median;
- how to interpret the mean when a quantitative variable is coded;
- how to describe the shape of a distribution (symmetry; skewness);
- how to present data (frequency tables; charts);
- what are weighted means and when to use them.

Data files contain a lot of information that must be summarized in order to be useful. If we look for instance at the variable **age** in the data file **GSS93 subset** that comes with the SPSS package, we will find 1500 entries, giving us the age of every individual in the sample. If we examine the ages of men and women separately, we cannot determine, by looking simply at the raw data, whether men of this sample tend to be older than women or whether it is the other way around. We would need to know, let us say, that the average age of men is 23 years and of women 20 years to make a comparison. The average is a descriptive measure.

Descriptive statistics aim at **describing** a situation by **summarizing** information in a way that highlights the important numerical features of the data. Some of the information is lost as a result. A good summary captures the essential aspects of the data and the most relevant ones. It summarizes it with the help of numbers, usually organized into tables, but also with the help of charts and graphs that give a visual representation of the distributions.

In this chapter, we will be looking at one variable at a time. Measures that concern one variable are called **univariate** measures. We will examine **bivariate** measures, those measures that concern two variables together, in Chapter 8.

There are three important types of univariate descriptive measures:

- measures of central tendency,
- measures of dispersion, and
- measures of position.

Measures of central tendency (sometimes called *measures of the center*) answer the question: What are the categories or numerical values that represent the *bulk* of the data in the best way? Such measures will be useful for comparing various groups within a population, or seeing whether a variable has changed over time. Measures of central tendency include the *mean* (which is the technical term for average), the *median*, and the *mode*.

Measures of dispersion answer the question: How spread out is the data? Is it mostly concentrated around the center, or spread out over a large range of values? Measures of dispersion include the standard deviation, the variance, the range (there are several variants of the range, such as the interquartile range) and the coefficient of variation.

Measures of position answer the question: How is one individual entry positioned with respect to all the others? Or how does one individual score on a variable in comparison with the others? If you want to know whether you are part of the top 5% of a math class, you must use a measure of position. Measures of position include percentiles, deciles, and quartiles.

Other measures. In addition to these measures, we can compute the *frequencies* of certain subgroups of the population, as well as certain *ratios* and *proportions* that help us compare their relative importance. This is particularly useful when the variable is qualitative, or when it is quantitative but its values have been grouped into categories.

The various descriptive measures that can be used in a specific situation depend on whether the variable is qualitative or quantitative. When the variable is quantitative, we can look at the *general shape of the distribution*, to see whether it is *symmetric* (that is, the values are distributed in a similar way on both sides of the center) or *skewed* (that is, lacking symmetry), and whether it is rather flat or rather peaked (a characteristic called *kurtosis*).

Finally, we can make use of charts to convey a visual impression of the distribution of the data. It is very easy to produce colorful outputs with any statistical software. It is important, however, to choose the *appropriate* chart, one that is *meaningful* and that *conveys* the most important properties of the data. This is not always easy, and you will have to pay attention to the way an appropriate chart is chosen, a choice that depends on the level of measurement of the variable.

It is very important to realize that the statistical measures used to describe the data pertaining to a variable depend on the level of measurement used. If a variable is measured at the nominal scale, you can compute certain measures and not others. Therefore you should pay attention to the *conditions* under which a measure could be used; otherwise you will end up computing numerical values that are meaningless.

Measures of Central Tendency

For Qualitative Variables

The best way to describe the data that corresponds to a qualitative variable is to show the **frequencies** of its various categories, which are a simple count of how many individuals fall into each category. You could then work out this count as a **percentage** of the total number of units in the sample. When you ask for the frequencies, SPSS automatically calculates the percentages as well, and it does it twice: the percentage with respect to the total number of people in the sample, and the percentage with respect to the valid answers only, called **valid percent** in the SPSS outputs. Let us say that the percentage of people who answered Yes to a question is 40% of the total. If only half the people had answered, this percentage would correspond to 80% of the valid answers. In other words, although 40% of the people answered Yes, they still constitute 80% of those who answered. SPSS gives you both percentages (the total percentage and the valid percentage) and you have to decide which one is more significant in a particular situation.

For instance, Table 3.1 summarizes the answers to a question about the legalization of marijuana, in a survey given to a sample of 1500 individuals.

Table 3.1 **A frequency table, showing the frequencies of the various categories, as well as the percentage and valid percentage they represent in the sample**

Should Marijuana Be Made Legal

		Frequency	Percent	Valid Percent
Valid	Legal	211	14.1	22.7
	Not legal	719	47.9	77.3
	Total valid	**930**	**62.0**	**100.0**
Missing		570	38.0	
Total		**1500**	**100.0**	

Table 3.1 tells us that the sample included 1500 individuals, but that we have the answers to that question for 930 individuals only. The percentage of positive answers can be calculated either out of the total number of people in the sample, giving 14.1% as shown in the Percent column, or out of the number of people for whom we

have answers, giving 22.7% as shown in the Valid Percent column. Which percentage is the most useful? It depends on the reason for the missing answers. If people did not answer because the question was asked of only a subset of the sample, the valid percentage is easier to interpret. But if 570 people abstained because they do not want to let their opinion be known, it is more difficult to interpret the resulting figures. A good analysis should include a discussion of the missing answers when their proportion is as important as it is in this example.

Table 3.1 comes from the SPSS output. When we write a statistical report, we do not include all the columns in that table. Most of the time, you would choose either the valid percentage (which is the preferred solution) or the total percentage, but rarely both, unless you want to discuss specifically the difference between these two percentages. The cumulative percentage is only used for ordinal or quantitative variables, and even then is included only if you plan to discuss it.

To describe the *center* of the distribution of a qualitative variable, you must determine which category includes the biggest concentration of data. This is called the mode. *The* **mode** *for a qualitative variable is the category that has the highest frequency* (sometimes called **modal category**).

The modal category could include more than 50% of the data. In this case we say that this category includes the **majority** of individuals. If the modal category includes less than 50% of the data, we say that it constitutes a **plurality**. We can illustrate this by the following situations concerning the votes in an election.

First situation:	Party A	54% of the votes
	Party B	21% of the votes
	Party C	25% of the votes,

Here we could say that Party A won the election with a *majority*. Compare with the following situation.

Second situation:	Party A	44% of the votes
	Party B	31% of the votes
	Party C	25% of the votes,

Here we can say that Party A won the election with a *plurality* of votes, but without a majority. If Parties B and C formed a coalition, they could defeat Party A. For this reason, some countries include in their electoral law a provision that, should the winning candidate or a winning party get less than the absolute majority of votes (50% + 1), a second turn should take place among those candidates who are at the top of the list, so as to end up with a winner having more than 50% of the votes.

A good description of the distribution of a qualitative variable should include a mention of the modal category, but it should also include a discussion of the pattern of the distribution of individuals across the various categories. Concrete examples will be given in the last section of this chapter.

For Quantitative Variables

Quantitative variables allow us a lot more possibilities. The most useful measures of central tendency are the mean and the median. We will also see how and when to use the mode. *The **mean** of a quantitative variable is defined as the sum of all entries divided by their number.*

In symbolic terms,

the mean of a *sample* is written as $\qquad \bar{x} = \dfrac{\sum x_i}{n}$, $\qquad$ and

the mean of a *population* is written as $\mu_x = \dfrac{\sum x_i}{N}$

These symbols are read as follows:

$\bar{x}$ $\quad$ is read as *x bar*, and it stands for the mean of a sample for variable X.

μ_x $\quad$ is read as *mu x*, and it stands for the mean of a population. The subscript x refers to the variable X.

x_i $\quad$ is read as *x i*. It refers to all the entries of your data that pertain to the variable X, which are labeled x_1, x_2, x_3, etc.

Σ $\quad$ is read as *sigma*. When followed by x_i, it means: add all the x_i's, letting i range over all possible values, that is, from 1 to n (for a sample) or from 1 to N (for a population).

n $\quad$ is the size of the sample, that is, the number of units that are in it.

N $\quad$ is the size of the population.

You may have noticed that we use different symbols for a population and for a sample, to indicate clearly whether we are talking about a population or a sample. We do not always need to write the subscript x in μ_x. We do it only when several variables are involved, and when we want to keep track of which of the variables we are talking about. In such a situation we would use μ_x, μ_y, and μ_z to refer to the mean of the population for the variables x, y, and z respectively. Notice that in the formula for the mean of a population, we have written a capital N to refer to the size of the population rather than the small n used for the size of a sample.

The mean is very useful to compare various populations, or to see how a variable evolves over time. But it can be very misleading if the population is not homogeneous. Imagine a group of five people whose hourly wages are: \$10, \$20, \$45, \$60 and \$65 an hour. The average hourly wage would be:

$$\bar{x} = \frac{10 + 20 + 45 + 60 + 65}{5} = \$40 \text{ an hour.}$$

But if the last participant was an international lawyer who charged \$400 an hour of consultancy, the average would have been \$107 an hour (you can compute it yourself), which is well above what four out of the five individuals make, and would be a misrepresentation of the center of the data.

In order to avoid this problem, we can compute the **trimmed mean**: you first eliminate the most extreme values, and then you compute the mean of the remaining ones. But you must indicate how much you have trimmed. In SPSS, one of the procedures produces a **5% trimmed mean**, which means that you disregard the 5% of the data that are farthest away from the center, and then you compute the mean of the remaining data entries.

The mean has a mathematical property that will be used later on. Starting from the definition of the mean, which states that $\bar{x} = \frac{\sum x_i}{n}$, we can conclude, by multiplying both sides by n, that:

$$\bar{x} * n = \sum x_i$$

In plain language, this states that the sum of all entries is equal to n times the mean.

We will discuss all the limitations and warnings concerning the mean in a later section on methodological issues.

THE MEAN OF DATA GROUPED INTO CLASSES

When we are given numerical data that is grouped into classes, and we do not know the exact value of every single entry, we can still compute the mean of the distribution by using the midpoint of every class. What we get is not the exact mean, but it is the closest guess of the mean that is available. If the classes are not too wide, the value obtained by using the midpoints is not that different from the value that would have resulted from the individual data.

Consider one of the intervals i with frequency f_i and midpoint x_i. The exact sum of all the entries in that class is not known, but we can approximate it using the midpoint. Thus, instead of the sum of the individual entries (not known) we will count the midpoint of the class f_i times. We obtain the following formula.

$$\text{Mean for grouped data} = \frac{\sum f_i * x_i}{n}$$

Here, n is the number of all entries in the sample. It is therefore equal to the sum of the class frequencies, that is, the sum of the number of individuals in the various classes. The formula can thus be rewritten as

$$\text{Mean for grouped data} = \frac{\sum f_i * x_i}{\sum f_i}$$

INTERPRETATION OF THE MEAN WHEN THE VARIABLE IS CODED

We often have data files where a quantitative variable is not given in its original form, but coded into a small number of categories. For instance, the variable Respondent's Income could be given in the form shown in Table 3.2.

Table 3.2 **Example of a quantitative variable that is coded into 21 categories, with a 22nd category for those who refused to answer**

Category	Code
Less than $1000	1
$1000–2999	2
$3000–3999	3
$4000–4999	4
$5000–5999	5
$6000–6999	6
$7000–7999	7
$8000–9999	8
$10,000–12,499	9
$12,500–14,999	10
$15,000–17,499	11
$17,500–19,999	12
$20,000–22,499	13
$22,500–24,999	14
$25,000–29,999	15
$30,000–34,999	16
$35,000–39,999	17
$40,000–49,999	18
$50,000–59,999	19
$60,000–74,999	20
$75,000 and more	21
Refused to answer	22

Thus, we would not know the exact income of a respondent. We would only know the category he or she falls into.

This kind of measuring scale poses a challenge. If we compute the mean with SPSS, we will not get the mean income. We will get the mean code, because it is the codes that are used to perform the computations. There is a data file that comes with SPSS where the income is coded in this way. This data file contains information about 1500 respondents, including information on the income bracket they fall into, coded as shown in Table 3.2. When we exclude the 22nd category, which consists of the people who refused to answer this question, the computation of the mean with SPSS produces the following result:

$$\text{Mean} = 12.35$$

What is the use of this number? It is not a dollar amount! If we look at Table 3.2, we see that the code 12 stands for an income of between $17,500 a year and $20,000 a year (with that last number excluded from the category). To interpret this number, we should first translate it into a dollar amount (it can be done with a simple rule). But even without transforming it into the dollar amount it corresponds to, we could use the mean code for comparisons. For instance, we will see in Lab 3 that if we compute the mean income separately for men and women, we get

Mean income for men: 13.9

Mean income for women: 10.9

(excluding the category of people who refused to answer).

Although the mean code does not tell us exactly the mean income for men and women, it still tells us that there is a big difference between men and women for that variable. Table 3.2 tells us that the code 13 corresponds to the income bracket $22,500–25,000, while the code 10 represents the income bracket $12,500–15,000. We can conclude that the difference in income between men and women, for that sample, is roughly around $10,000 a year.

We see that that when the variables are coded, the interpretation of the mean requires us to translate the value obtained into what it stands for. For quantitative variables coded this way, it may also be useful to find the frequencies of the various categories, as we did for nominal variables. For the example at hand, we would get Table 3.3 as shown.

The conclusion of the preceding discussion is that when we have an ordinal variable with few categories, or even a quantitative variable that has been recoded into a small number of categories, it may be useful to compute the frequency table of the various categories, in addition to the mean and other descriptive measures.

Weighted Means

Consider the following situation: you want to find the average grade in an exam for two classes of students. The first class averaged 40 out of 50 in the exam, and the second class averaged 46 out of 50. If you put the two classes together, you *cannot* conclude that the average is 43. This is so because the classes may have different numbers of students. Suppose the first class has 20 students, and the second one 40 students. In other words, we have the data shown in Table 3.4.

To compute the average grade for the two classes taken together, we do not need to know the individual scores of each student. Indeed, we have seen before that a sum of n scores is equal to its average times n. We will use this to obtain the formula shown below for weighted means.

The mean for the two classes taken together can be written as

Table 3.3 **Frequencies of the various income categories for the variable Income**

Respondent's income

	Frequency	Valid Percent
LT $1000	26	2.6
$1000–2999	36	3.6
$3000–3999	30	3.0
$4000–4999	24	2.4
$5000–5999	23	2.3
$6000–6999	23	2.3
$7000–7999	15	1.5
$8000–9999	31	3.1
$10,000–12,499	55	5.5
$12,500–14,999	54	5.4
$15,000–17,499	64	6.4
$17,500–19,999	58	5.8
$20,000–22,499	55	5.5
$22,500–24,999	61	6.1
$25,000–29,999	84	8.5
$30,000–34,999	83	8.4
$35,000–39,999	54	5.4
$40,000–49,999	66	6.6
$50,000–59,999	38	3.8
$60,000–74,999	23	2.3
$75,000+	44	4.4
Refused to answer	47	4.7
Total	994	100.0
Missing	506	
Grand Total	1500	

Table 3.4 **Two classes of different size and the mean grade in each**

	Average Grade out of 50	Number of Students
Class A	40	20
Class B	46	40

$$\frac{\text{Sum of all scores in class A + Sum of all scores in class B}}{60}$$

The sum of all scores in class A can be replaced by the average score (40) times 20, since there are 20 students in this class. And the sum of all scores in class B can be replaced also by its average score (46) times 40, since this class includes 40 students. The equation for the mean becomes:

$$\frac{(40 \times 20)}{60} + \frac{(46 \times 40)}{60}$$

This can now be written as:

$$\text{mean of the two classes combined} = 40 \times (\mathbf{20/60}) + 46 \times (\mathbf{40/60})$$

or again as:

$$\text{mean of the two classes combined} = 40 \times (\mathbf{1/3}) + 46 \times (\mathbf{2/3})$$

The last formula is important: we see that the average grade of class A is multiplied by the **weight** of class A, which is its relative importance in the total population. Class A forms 1/3 of the total population (20 students out of 60) and class B 2/3 of the total (40 students out of 60). The underlying formula is:

$$\text{Average grade for the two classes: } 40 \times w_1 + 46 \times w_2$$

The $\mathbf{w_i}$'s are called the **weights** of the various classes. In this case, the weight is an expression of the number of people in each class compared to the total population of the two classes.

The general formula is as follows.

If you have n values	$x_1, x_2, x_3, \ldots$ etc.,
each having the corresponding weights:	$w_1, w_2, w_3, \ldots$ etc.,
the **weighted mean** is given by	$x_1 w_1 + x_2 w_2 + x_3 w_3 + \cdots + x_n w_n$

The weights are positive numbers and must add up to 1. That is:

$$w_1 + w_2 + w_3 + \cdots + w_n = 1.$$

The weights are not always a reflection of the size of the various groups involved. If you are computing the weighted average of your grades during your college studies, the weights could be proportional to the credits given to each course, or they could be an expression of the importance of the course in a given program of studies. A Faculty of Medicine may weight the grades of its candidates by giving a bigger weight to Chemistry and Biology than Art History, for instance.

Example

A buyer wants to evaluate several houses she has seen. She attributes a score out of ten to each house on each of the following items: size, location, internal design, and quality of construction. Any house having a score less than 5 on any item would not be acceptable. The resulting scores for three houses that are seen as acceptable on all grounds are recorded in Table 3.5. The buyer does not attribute the same importance to each item. The size of the house is the most important quality. The quality of the construction is also very important, but not as important. The buyer attributes a weight to each item, which reflects the importance of that item for her. The weights are given in the last column.

Table 3.5 **Scores given to three houses on four items, and their weights**

Item	House A	House B	House C	Weight of item
Size	9	7	6	0.4
Location	5	9	10	0.1
Internal design	6	5	8	0.2
Quality of construction	7	9	7	0.3

We can now calculate the weighted average score for each house, using the formula for weighted means given above.

For house A: weighted mean score: $10 \times 0.4 + 5 \times 0.1 + 6 \times 0.2 + 7 \times 0.3 = 7.8$
For house B: weighted mean score: $7 \times 0.4 + 9 \times 0.1 + 5 \times 0.2 + 9 \times 0.3 = 7.4$
For house C: weighted mean score: $6 \times 0.4 + 10 \times 0.1 + 8 \times 0.2 + 7 \times 0.3 = 7.1$

We see that house A obtained the highest weighted score. The total, unweighted score of house C is higher than that of house A. But because the items do not all have the same importance, house A ended up having a higher weighted score.

THE MEDIAN AND THE MODE

The **median** is another measure of central tendency for quantitative variables. It is defined as the value that sits right in the middle of all data entries when they are listed in ascending order. If the number of entries is odd, there will be one data entry right in the middle. If the number of entries is even, we will have *two* data entries in the middle, and the median in this case will be their average. Here are two examples.

Case 1: variable X 2, 3, 4, 4, 5, 5, **5**, 6, 7, 8, 11, 13, 13
Case 2: variable Y 2, 3, 4, 4, 5, **5**, **6**, 7, 8, 11, 13, 13

For the variable X we have 13 entries. The value 5 sits in the middle, with six entries equal or smaller than it, and six entries equal or larger. The median for X is thus 5. But for variable Y, we have 12 entries. There are therefore two entries in the middle of the ordered list, not just one. The median will be the average of the two, that is $(5 + 6) \div 2 = \textbf{5.5}$.

The median is not sensitive to extreme values. Suppose, for instance, that the entries for variable X were: 2, 3, 4, 4, 5, 5, 5, 6, 7, 8, 11, 13, 60. Although the last entry is very large compared to the others, it does not affect the median, which is still 5. The mean, however, would have been affected (compute it yourself for the two situations and see how different it would be). For this reason, the median is a better representative of the center when there are extremely large values on one side of it. But the mean is more useful for statistical computations, as we will see in the coming sections.

Half the population has a score that is lower than or equal to the median, and the other half has a score larger than the median or equal to it. This way of formulating the median is very useful in situations where the distribution is skewed (such as the distribution of income) or in situations where time is involved, especially when processes have not been completed by everybody, as illustrated below.

Examples of the use of the median

- We are told that the average age at first marriage for a population is 22 years for women, and 25 for men. The median for women is 21, and for men it is 24. This means that by the time they reached 21 years of age, half the women in this population were married. For men, half of them were married by the age of 24.
- In a research on the time taken by immigrants to find a job, 500 new immigrants who arrived at least three years ago are interviewed. The mean can not be found because some of them have not found a regular or full-time job yet. But it is found that the median time taken for them to find a regular, full-time job was 18 months for men, and 5 months for women. This means that by the 18th month after arrival, 50% of the men had found a job. Women were faster in finding regular full-time jobs: 50% had a job within 5 months of their date of arrival.

Because the median involves only the *ordered* list of data entries, it can be used if the quantitative variable is measured at the ordinal level. But if the number of categories is small, the median is not very useful.

The **mode** can also be used for quantitative variables. When the values are grouped into classes, the mode is defined as it is for qualitative variables: it is the class that has the highest frequency. But the mean and median remain the best descriptive measures for quantitative variables. If the variable is continuous and the values have not been grouped into classes, the mode is the value at which a *peak* occurs in the graph representing the distribution.

COMPARISON OF THE MEAN AND THE MEDIAN
Both the mean and the median are measures of central tendency of a distribution, that is, they give us a central value around which the other values are found. They are therefore very useful for comparing different samples, or different populations,

or samples with a population, or a given population at different moments in time to see how it has evolved. However, each of the mean and the median has its advantages and its drawbacks.

The mean takes into account every single value that occurs in the data. Therefore, it is sensitive to every value. A single very large value can boost the mean up if the number of entries is not very large. For instance, if one worker in a group of 20 workers won a $1 million lottery ticket, the average wealth of those 20 would look artificially high. The median is not sensitive to every single value. In a distribution where the largest value is changed from 60 to 600, the median would not change. The mean would.

It follows from these remarks that the mean is a more sophisticated measure, because it takes every value into account. Indeed, it is the mean that is used to compute the standard deviation, which a measure of dispersion that will be seen below. However, in situations where the distribution is not very symmetric, and where there are some extreme values on only one side of the distribution, the mean will tend to be shifted towards the extreme values, whereas the median will stay close to the bulk of the data. Therefore, whenever the distribution is highly skewed, the median is a better representative of the center of the distribution than the mean. This is true for variables such as income or wealth, where the distribution among individuals in a country, and also worldwide, is highly skewed. For such a variable, the median is a more accurate representative of the central tendency of the distribution.

Measures of Dispersion

For Qualitative Variables

There are not many measures of dispersion for qualitative variables. One of the measures we can compute is the **variation ratio**. It tells us whether a large proportion of data is concentrated in the modal category, or whether it is spread out over the other categories. The variation ratio is defined as

$$\text{variation ratio} = \frac{\text{number of entries not in the modal class}}{\text{total number of entries}}$$

It is a positive number smaller than one. If this ratio is close to zero, it indicates a great homogeneity, almost every unit being in the modal class. The farther it is from zero, the greater the dispersion of the data over the other categories. Like many other measures, this one is easy to interpret when doing comparisons. For instance, if we compare the sizes of the various linguistic groups in two cities where several languages are spoken, we can use the variation ratio to assess the degree of heterogeneity in each city. Here is an example.

City	Linguistic groups	Percentage
City A	French speaking	30%
	English speaking	34%
	Chinese speaking	20%
	Other	16%
	Total	**100%**
City B	French speaking	28%
	English speaking	40%
	Chinese speaking	20%
	Other	12%
	Total	**100%**

The variation ratio for city A would be $(30 + 20 + 16)/100 = 0.66$, and for city B it would be $(28 + 20 + 12)/100 = 0.60$, showing that city A is a little more heterogeneous than city B.

For Quantitative Variables

There are many ways of measuring the dispersion for quantitative variables. The simplest is the range, but we also have various forms of restricted range, we have the deviation from the mean, the standard deviation, the variance and finally the coefficient of variation. Let us go through these measures one at a time.

RANGE

The **range** is the simplest way of measuring how spread out the data is. You simply subtract the smaller entry from the larger one and add 1, and this tells you the size of the interval over which the data is spread out. For example, you would describe a range of values for the variable **Age** as follows:

> In this sample, the youngest person is 16 years old and the oldest 89, spanning a range of 74 years $(89 - 16 + 1)$.

But we may have extreme values that give a misleading impression about the dispersion of the data. For instance, suppose that a retired person decided to enroll in one of our classes. We could then say that the ages of the students in this class range from 16 years up to 69 years, but that would be misleading, as the great majority of students are somewhere between 17 years old and maybe 23 or 24 years old. For this reason, we can introduce variants of the notion of range.

The C_{10-90} **range**, for instance, computes the range of values after we have dropped 10% of the data at each end: the 10% largest entries and the 10% smallest entries. This statistic gives us the range of the remaining 80% of data entries. We can also compute the **5% trimmed range** by deleting from the computation the 5% of values that are the farthest away from the mean. We will also see in a forthcoming section something called a *box-plot*, that shows us graphically both the full range, and the range of the central 50% of the data after you have disregarded the top 25% and the bottom 25%. This last range is called the **interquartile range**, the distance between the first and third **quartiles**, which are the values that split the data into four equal parts.

These various notions of the range do not use the exact values of *all* the data in their computation. The following measures do.

STANDARD DEVIATION

The most important measure is the standard deviation. To explain what it is we must first define some simpler notions such as the deviation from the mean. For an individual data entry x_i the **deviation from the mean** is the *distance* that separates it from the mean. If we want to write it in symbols, we will have to use two different symbols, depending whether we have a sample or a population.

For a sample, the deviation from the mean is written: $(x_i - \overline{x})$
For a population, the deviation from the mean is written: $(x_i - \mu)$

The list of all deviations of the mean may give us a good impression of how spread out the data is.

Example

Consider the following distribution, representing the grades out of ten of a group of 14 students:

$$4, 5, 5, 6, 7, 7, 8, 8, 8, 9, 9, 9, 10, 10$$

Here the mean is given by $105/14 = 7.5$. The deviations from the mean are given in Table 3.6.

But that list may be long. We want to summarize it, and end up with a single numerical value that constitutes a measure of how dispersed the data is. We could take the mean of all these deviations. If you perform the computation for the mean deviation, you will get a mean deviation equal to zero (do the computation yourself on the preceding example). This is no accident. Indeed, we can easily show that the mean of these deviations is necessarily zero, as the positive deviations are cancelled out by the negative deviations.

Table 3.6 **Calculation of the deviations from the mean**

Data entry x_i	Deviation from the mean: $(x_i - \bar{x})$
4	$4 - 7.5 = -3.5$
5	$5 - 7.5 = -2.5$
5	$5 - 7.5 = -2.5$
6	$6 - 7.5 = -1.5$
7	$7 - 7.5 = -0.5$
7	$7 - 7.5 = -0.5$
8	$8 - 7.5 = 0.5$
8	$8 - 7.5 = 0.5$
8	$8 - 7.5 = 0.5$
9	$9 - 7.5 = 1.5$
9	$9 - 7.5 = 1.5$
9	$9 - 7.5 = 1.5$
10	$10 - 7.5 = 2.5$
10	$10 - 7.5 = 2.5$

The mathematical proof (which is given only for those who are interested and which can be ignored otherwise) goes like this:

Sum of all deviations from the mean =

$$\sum (x_i - \bar{x}) = \sum x_i - \sum \bar{x} = n * \bar{x} - n * \bar{x} = 0$$

(Explanation: Recall that the sum of all entries is equal to n times the mean, and that the mean, in the second summation, is counted n times. This is why we get n times the mean twice, once with a positive sign, and once with a negative sign.)

We thus conclude that the deviations from the mean always add up to zero, and therefore we cannot summarize them by finding their mean. The way around this difficulty is the following: we will square the deviations, and then take their mean. By squaring the deviations, we get rid of the negative signs, and the positive and negative deviations do not cancel out any more. This operation changes their magnitude, however, and gives an erroneous impression about the real dispersion of data, since the deviations are all squared. This distortion will be corrected by taking the square root of the result, which brings it back to an order of magnitude similar to the original deviations. In summary, we end up with the following calculation:

Standard deviation for a population, denoted by the symbol σ

$$\sigma = \sqrt{\frac{\sum (x_i - \mu)^2}{N}}$$

In the case of a sample, μ will be replaced by $\bar{x}$ and N will be replaced not by n, but by $n - 1$. The reason why we write $n - 1$ instead of n is due to some of the mathematical properties of the standard deviation. It can be proven that using $n - 1$ in the formula gives a better prediction of the standard deviation of a population when we know that of the sample.

Conclusion: the **standard deviation for a sample**, denoted by the symbol **s**, is given by:

$$s = \sqrt{\frac{\sum (x_i - \bar{x})^2}{n - 1}}$$

The standard deviation (often written **st.dev.**) is the most powerful measure of dispersion for quantitative data. It will permit us to do very sophisticated descriptions of various distributions. All the calculations of statistical inference are also made possible by the use of the standard deviation.

VARIANCE

Another useful measure is the **variance**, which is defined as the square of the standard deviation. It is thus given by

$$\text{variance of a sample} = s^2$$

or

$$\text{variance of a population} = \sigma^2$$

THE COEFFICIENT OF VARIATION

Finally, we can define the coefficient of variation. To explain the use of this measure, suppose you have two distributions having the means and standard deviations given below:

Distribution 1	mean = 30	st. dev. = 3
Distribution 2	mean = 150	st. dev. = 3

In one case the center of the distribution is 30, indicating that the data entries fall in a certain range *around* the value 30. Their magnitude is around 30. In the other case, the mean is 150, indicating that the data entries fall in a range around the value 150 and have an average magnitude of 150. Although they have the same dispersion (measured by the standard deviation), the relative importance of the dispersion is not the same in the two cases because the magnitude of the data is different. In one case the entries revolve around the value 30, and the standard deviation is equal to 10% of the average value of the entries. In the other case, the entries revolve around the value 150 and the standard deviation is about 3/150, that is, 2% of the average value of the entries, a value which denotes a smaller relative variation.

There is a way to assess the relative importance of the variation among the entries, by comparing this variation with the mean. The measure is called the *coefficient of variation*. The **coefficient of variation** is defined as the standard deviation divided by the mean, and multiplied by 100 to turn it into a percentage. The formula is thus:

$$\text{Coefficient of variation } CV = \frac{\sigma}{\mu} \times 100$$

This measure will only be used occasionally.

Measures of Position

Measures of position are used for quantitative variables, measured at the numerical scale level. They could sometimes be used for variables measured at the ordinal level. They provide us with a way of determining how one individual entry compares with all the others.

The simplest measure of position is the quartile. If you list your entries in an ascending order according to size, *the **quartiles** are the values that split the ranked population into four equal groups.* Twenty-five percent of the population has a score less or equal than the 1st quartile (Q_1), 50% has a score less than the 2nd quartile (Q_2), and 75% has a score less than the 3rd quartile (Q_3). Recall that we have seen earlier a measure of dispersion called the interquartile range, which is the difference between Q_1 and Q_3. Figure 3.1 illustrates the way the quartiles divide the ordered list of units in a sample or in a population.

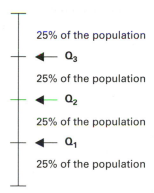

Figure 3.1 **The quartiles are obtained by ordering the individuals in the population by increasing rank, and then splitting it into four equal parts. The quartiles are the values that separate these four parts**

In a similar way, we can define the **deciles**: they split the ranked population into ten equal groups. If a data entry falls in the first decile it means that its score is among the lowest 10%. If it is in the 10th decile it means it is among the top 10%.

The most common measure of position, however, is the *percentile rank*. The data is arranged by order of size (recall it must be quantitative) and divided into 100 equal groups. The numerical values that separate these 100 groups are called **percentiles**. The **percentile rank** of a data entry is the rank of the percentile group this entry falls into. For example, if you are told that your percentile rank in a national exam is 83, this means that you fall within the 83rd percentile. Your grade is just above that of 82% of the population, and just below that of 17% of the population. You will learn in the SPSS session how to display the percentile ranks of the data entries.

You may have realized by now the connection between the median and the various measures of position, since the median divides your ranked population into two equal groups. The median is equal to the 50th percentile. It is also equal to the 5th decile, and of course the 2nd quartile.

Other Measures

We have seen above that when a variable is measured by a nominal scale we cannot compute measures such as the mean or the standard deviation. We can just count how many units are in each of the categories of the variable. If the variable is Respondent's sex, for example, we can say something such as 'There are 40 men in this sample.' However, a number like that is rarely meaningful by itself. It becomes more meaningful when you compare it to something else: to the number of women, for instance, or to the total number of people in the sample. You can then come up with statements such as:

'60% of the people in the sample are women'

or

'There are twice as many men as there are women in this population'

The first statement is an example of a **percentage**, and the second of a **ratio**. We could also have **proportions**. Let us examine these measures briefly.

Ratios

Ratios are the result of the division of the frequency of one category by the frequency of another category. If we have a sample consisting of 40 men and 20 women, we would say that the ratio of men to women is 40 to 20. If we divide 40 by 20, we get a ratio of 2 to 1, which means that there are 2 men for every woman. If we had 16 men and 20 men, we would say the ratio is $16 \div 20 = 0.8$, which means that there are 0.8 men for every woman.

We can do such computations whenever we have categories. Even if the variable is quantitative and it has been recoded into categories, we can find the ratio of one category compared to any of the other ones.

In general, then, a ratio could involve decimals. But it is more convenient to formulate it as whole numbers. One way of doing it is the following.

Suppose you have a ratio of men to women which is equal to 0.73. It means that there are 0.73 men for every woman. You can write it as:

$$0.73:1$$

We will now multiply both numbers by 2, then by 3, then by 4, etc., until we get the first number to be close to an integer. The ratio is not changed by this operation since we are multiplying both sides by the same constant. Here is what we get.

Multiply by	2	and get	1.46:2
Multiply by	3	and get	2.19:3
Multiply by	4	and get	2.92:4
Multiply by	5	and get	3.65:5

The third line is the closest to an integer, so this is the one we will keep and say:

There are about 3 men for every 4 women.

This way of expressing ratios is more intuitive and it conveys a sense of the relative size of the groups that is easier to grasp.

Percentages and Proportions

Percentages and proportions compare the frequency of a given category not to another one, but to the *totality* of the sample. Percentages and proportions are calculated in almost the same way, the only difference being that the proportion is out of 1 and the percentage is out of 100. Move the decimal point by two places to the right and you transform a proportion into a percentage. The valid percent compares the frequency of the category to the number of valid answers, which means that all the missing answers are excluded from the computation.

Graphical Representation of the Distribution of Data

The measures we have seen are very good for describing the data accurately. But it is also desirable to convey a visual impression of the distribution of the data. Graphs and charts allow us to see *patterns* that cannot easily be conveyed with words or numbers alone.

In the second SPSS Lab on descriptive statistics (Lab 4), you will learn how to use the SPSS Help menu to familiarize yourself with the various sorts of charts and how to interpret them. Here, we will just give basic definitions.

The kind of chart chosen to represent a given variable will depend on the level of measurement of the variable. A chart (any chart) must always include a title that explains what it is about, and a legend that indicates how to interpret the different elements of the chart. The variables that are represented on the axes must be clearly identified, together with the units used to measure them. Fortunately, all this information is automatically included with the SPSS charts. We will now examine the basic types of charts and discuss briefly when they are most useful.

Bar Charts

They are *simple*, *clustered* or *stacked*.

A **simple bar chart** consists of some columns or bars. Each column represents a quantity or a percentage, and it corresponds to one category of the variable studied. For instance, the X-axis could be the variable Marital Status, with categories: married, widowed, etc. Each bar represents the frequency (either count or percentage) of the corresponding category.

In general, the variable on the X-axis is a qualitative variable. For instance, it could be the variable Mother Tongue. Every bar would then represent the number of

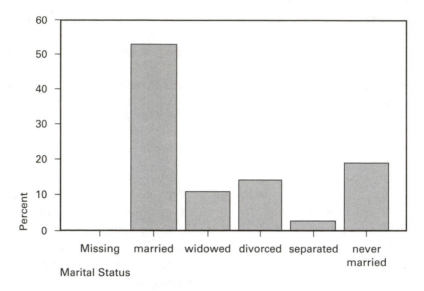

Figure 3.2   **A bar chart representing the size (in percentage) of the various categories for the variable Marital Status**

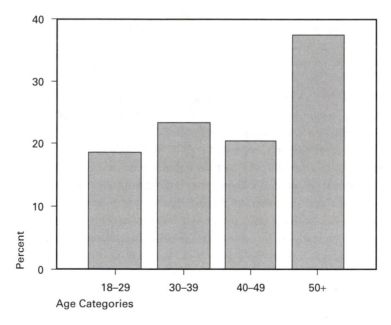

Figure 3.3   **A bar chart representing the various age categories in percentages of the whole sample**

people who speak a given language. You can choose to have the Y-axis represent *percentages* instead of counts. The chart shown in Figure 3.2 represents the percentages of the various marital categories.

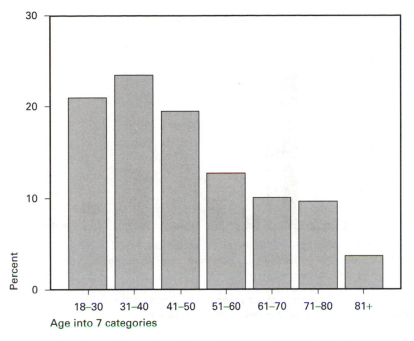

Figure 3.4 **A bar chart where the category 50+ years has been broken down into four categories**

The variable on the X-axis could also be a *quantitative variable that has been grouped into a small number of categories*. For instance, we could have **agecat4** as the variable on the X-axis. The bars would then represent the number of people found in each of the four age categories. In this kind of bar graph, you must be careful about the *range* (that is, the length of the interval) of each of the categories. If the categories are intervals that do not have the same length, you may get the wrong impression that one group is more numerous than the other, such as with the group of people who are 50 years old or more in the chart shown in Figure 3.3.

However, this group (50 years and older) spans a range of ages which is much wider than the other groups: close to 40 years (from 50 years to 89 years exactly). If we regroup the respondents into age categories that are equal or almost equal, we get the chart in Figure 3.4.

This bar chart is a much better representation of the distribution of ages than the previous one.

In a **clustered bar chart**, each column is subdivided in several columns representing the categories of a second variable. For instance, each column could be split in two, for men and for women. Figure 3.5 provides an example of a clustered bar chart where the height of the columns represents the number of people in each category.

In a clustered bar chart, it is generally preferable to display the percentages of the various categories rather than their frequencies. Look for instance at the clustered bar chart displayed in Figure 3.5. We see that in every category, women are more

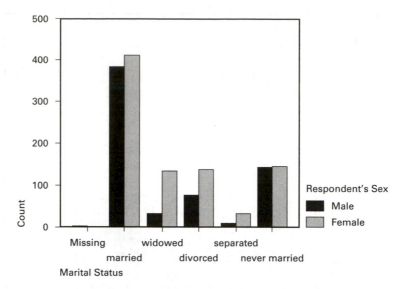

Figure 3.5   **A clustered bar chart where the height of the columns represents the number of people in each category**

numerous than men. This is so because the sample as a whole contains more women. This chart does not allow us to assess how the percentages of men and women compare in each category. If we display the percentages rather than the frequencies (the count), we get the chart illustrated in Figure 3.6.

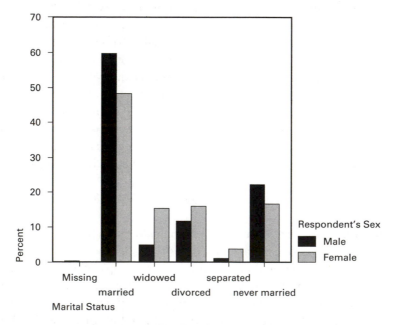

Figure 3.6   **A clustered bar chart displaying the percentages rather than the frequencies**

We see now that percentage-wise, there are a lot more women who are widows than men who are widowers. In that sample, it also happens that the divorced women are slightly more numerous than the divorced men (divorced women whose ex-husband has died are not counted in the Widow category but in the Divorced category). Although the sample used here is not necessarily representative of the whole American population, it does illustrate a social reality: as in many other societies, women tend to live longer than men. Therefore, the percentage of women in the categories Widowed and Divorced is larger than the percentage of men, and consequently lower than the percentage of men in all other categories, even if their numbers are bigger.

In a stacked bar chart, rather than being adjacent, the split columns are stacked one on top of the other, as shown in the Figure 3.7.

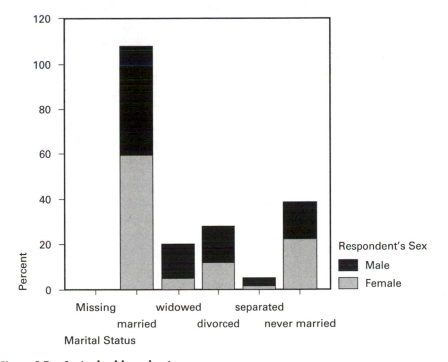

Figure 3.7 **A stacked bar chart**

The advantage of a stacked bar chart, as opposed to a clustered bar chart, is that it shows the overall importance of the categories (married, widowed, etc.), while at the same time showing how they are broken down into the categories of another variable such as Sex.

Bar charts are most adequate when you want to highlight the *quantity* associated with every category on the X-axis. A bar chart where the vertical axis does not start at 0 can be very misleading, for if the columns are truncated at their base, the differences in height between them can appear to be more important than they really are. Consequently, as a general rule, bar charts should start at zero and should not be truncated from their base.

Finally, it should be said that bar charts could also be presented horizontally, by interchanging the X- and Y-axes.

Pie Charts

Pie charts (Figure 3.8) are most useful when you want to illustrate proportions, rather than actual quantities. They show the relative importance of the various categories of the variable. In SPSS you have the option of including missing values as a slice in the pie, or excluding them and dividing the pie among valid answers. The details of how to do that are explained in Lab 5. Pie charts are better suited when we want to convey the way a fixed amount of resources is allocated among various uses. For instance, the way a budget is spent over various categories of items is best represented by a pie chart. When the emphasis is on the amount of money spent on each budget item, rather than on the way the budget is allocated, a bar chart is more suggestive. However, both bar charts and pie charts are appropriate to represent the distribution of a nominal variable, and there is no clear-cut line of demarcation that would tell us which of the two is preferable.

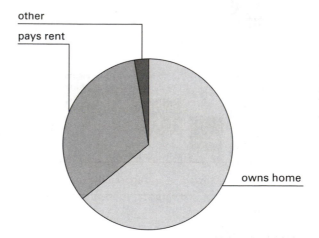

Figure 3.8   **Pie chart illustrating the proportion of people who own a home as compared to those who pay rent. One of the options in the pie chart command allows you to either include or exclude the category of missing answers. In this diagram it has been excluded from the graph**

Histograms

Histograms are useful when the variable is *quantitative*. The data are usually grouped into classes, or intervals, and then the frequency of each class is represented

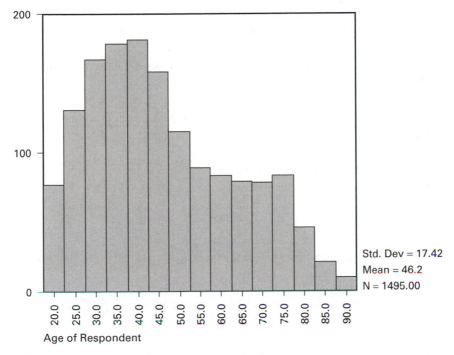

Figure 3.9   **Illustration of the histogram for the variable Age**

by a bar. The bars in a histogram are adjacent, and not separated as in a bar chart, because the numerical values are continuously increasing. For instance, if you draw the histogram of the variable Respondent's Age (Figure 3.9), you will see the pattern of the distribution of the individuals of the sample across the various categories. Contrary to a bar chart, which is used for a qualitative variable, the columns of the histogram cannot be switched around. You can switch around the categories of a variable measured at the nominal level, but not those of an ordinal or quantitative variable.

When producing a histogram with SPSS, the program automatically selects the number of classes (usually no more than 15) and divides the range of values accordingly into intervals of equal size. In the histogram shown in Figure 3.9, the **midpoints** of the classes are shown on the graph. They are:

20, 25, 30, 35, etc.

Therefore, the **class limits** (that is, the cut-point between one class and the next) are the values in between: 22.5, 27.5, 32.5, etc. We can infer that the lower limit of the first class is 17.5 years, and the upper limit of the last class is 92.5 years.

DRAWING A HISTOGRAM MANUALLY

In order to draw a histogram manually, you must go through the following steps.

1. Determine the range of values, that is, the difference between the smallest and largest values.
2. Determine the number of classes you wish to have. Usually, a histogram would have anywhere between 5 and 15 classes, but this is not a rigid rule. The number of classes will depend on the degree of detail you wish to see in your diagram.
3. Divide the range of values by the number of classes to get the length of the class interval. Round it up or down to make the calculation easy. The length of the class interval should be a whole number, and if possible a multiple of 5 or a multiple of 10.
4. Start your histogram at a whole number that is smaller than or equal to the smallest value, in such a way as to make the computation of the class interval easy.

Example

Suppose you have the ages of a sample of 200 people, the youngest being 18 and the oldest 67, and the ages being written as whole numbers without decimals. That makes a range of $67 - 18 + 1 = 50$ years. With this range of 50 years, we could have either 10 or 11 classes spanning 5 years each, depending on where we start. For example, we can draw a histogram that spans the ages from 15 to 70 in the following way:

Class	Class limits	Midpoint
$15 \leq x \leq 19$	15 and 19	17
$20 \leq x \leq 24$	20 and 24	22
etc...		
$65 \leq x \leq 69$	65 and 69	67

We could also use the values 20, 25, 30, etc. as midpoints, which would give the following organization of the data.

Class	Class limits	Midpoint
$18 \leq x \leq 22$	18 and 22	20
$23 \leq x \leq 27$	23 and 27	25
etc...		
$63 \leq x \leq 67$	63 and 67	65

The subtle issues that are related to the determination of the class limits and of the class intervals in situations where the values are continuous vs. discrete, and which were part of any standard textbook of descriptive statistics, are less relevant now as most statistical software would make these computations automatically.

Histograms are drawn in such a way that they have the following property: if you compare the surface of any bar to the total surface of the whole histogram, you should get exactly the percentage of data represented by the bar. When the data is quantitative and ungrouped, SPSS groups it in class intervals automatically, with all classes equal in length. If the grouping has been done into unequal classes, it is better to use bar charts rather than histograms, because the pattern of the distribution is less relevant than the number of individuals in the various classes.

Finally, we want to point out that the number of classes in a histogram could well be greater than 15. For instance, the demographic profile of the population of a country is often represented by two histograms (for males and females) drawn back

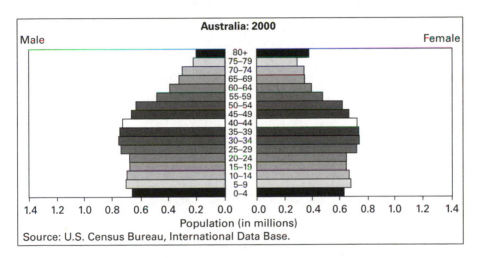

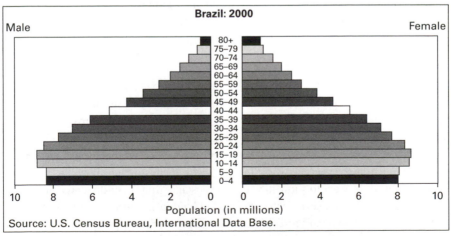

Figure 3.10 **Example of a pyramid of ages in a developing country (Brazil) and in a developed country (Australia). The demographic structure of the population is immediately appearent. (Source: US Census Bureau)**

to back with horizontal bars pointing to opposite directions, where each bar represents a five-year span. This type of histogram is called a **population pyramid**. In a population pyramid, the last class is left open. Usually it is the '80 years and more' class, as shown in Figure 3.10.

FREQUENCY POLYGONS AND DENSITY CURVES

If we join all the midpoints at the top of the columns in a histogram, we get what is called a **frequency polygon**. The polygon shows the general pattern of the distribution. Imagine now a frequency polygon drawn on a histogram with very large number of columns. We could redraw it as a smooth curve, called a **density curve** (Figure 3.11). A density curve is drawn in such a way that its surface is equal to 1. And if we look at the surface under the curve between any two values, it tells us the exact proportion of data that falls within these two values. We can now be more specific about the definition of the mode for a quantitative variable.

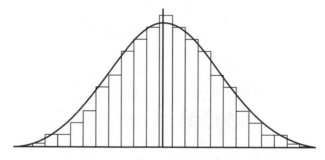

Figure 3.11   **A density curve can be thought of as the curve resulting from joining the midpoints at the top of the various bars of a histogram with a large number of classes**

If the variable is represented by a histogram, the **mode** is *the class with the highest frequency*. If it is represented by a density curve, the mode is *the x-value that corresponds to the highest point on the density curve*.

HISTOGRAM OR BAR CHART?

When we have a quantitative variable that has been grouped into a small number of categories, we can represent it either by a histogram or by a bar chart. But which of the two representations is better? It depends on what we want to convey. To explain this point, consider a situation where we have the variable Age represented by seven categories as shown in Figure 3.4. If we want to convey how the ages of the sample studied are distributed over the whole range of ages, the histogram shown in Figure 3.9 is better. But if we want to show how the various age groups are divided among men and women, or among married vs. unmarried individuals, a clustered bar chart allows us to do that, as shown in Figure 3.12. A histogram would not permit us to juxtapose corresponding categories of age groups for men and women.

In Figure 3.12, we see the distribution separately for men and women, and we can determine that women are more represented in the older categories, as they tend to live

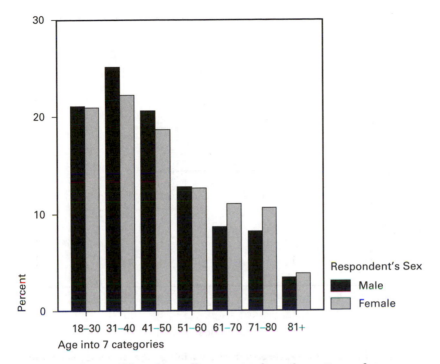

Figure 3.12 **A clustered bar chart allows us to show the pattern of ages separately for men and for women. It is appropriate for a quantitative variable grouped into a small number of categories**

longer than men. Notice that the vertical axis represents the percentages, not the frequencies. If we made it represent the frequencies instead, we would still get the same general shape, but we would not be able to determine whether men or women are more represented in a given class, as the overall number of women in this sample is greater than the overall number of men. In almost every age category, we would therefore find more women, not because a higher percentage of women (as opposed to men) fall into that category, but because there are more women in the sample as a whole.

Box plots

Box plots are very useful to show how the values of a *quantitative* variable are distributed. The box plot indicates the minimum and maximum values, and the three quartiles. The central 50% of the data (the 2nd and 3rd quarters) are represented as a shaded solid box, whereas the first and last quarters are represented by thin lines.

The box plot gives automatically the *five-number summary* of the data: the minimum, the 1st quartile, the median (which is the 2nd quartile), the 3rd quartile, and the maximum.

In symbols the **five-number summary** is given by: **Min, Q_1, Median, Q_3, Max**. The box plot is shown in Figure 3.13.

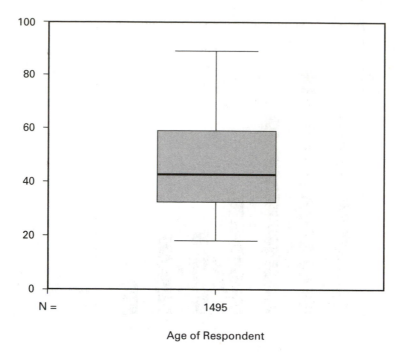

Age of Respondent

Figure 3.13   **The box plot representing the variable Age of respondent. We can read off directly the five-number summary of the distribution**

Box plots can also be used to represent several similar variables on the same graph, allowing comparisons. You could also split a population into several separate groups (such as men and women) and have a separate box plot for each group, drawn in the same graph, next to each other, to permit comparisons. This is illustrated in Figure 3.14 where five box plots of respondents' income are drawn, for the various groups defined by educational level. This figure illustrates clearly how the income varies with the highest level of education attained. We should note, however, that this data comes from a file where the income is not measured as a continuous scale variable, but is coded into 21 categories, and that the 22nd category is made up, as explained earlier in this chapter. More details are found in Lab 5.

Line Charts

Line charts are most useful to represent the variation of a quantitative variable over time. The X-axis represents the time line, and the Y-axis represents some quantitative variable. For example, the variable could be the number of students enrolled in a given program, or the inflation rate, or the market value of a given portfolio of stocks. The line chart would show how the variable increases or decreases as time goes by. A common mistake sometimes made intentionally consists in not showing

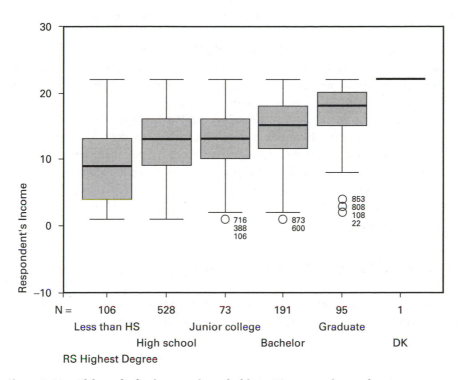

Figure 3.14 **Although the income is coded into 22 categories and not given as a dollar amount, the comparisons of the income for each educational level gives us a good idea of how incomes vary as a function of education**

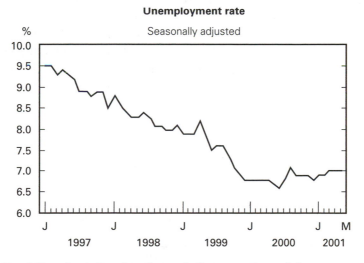

Figure 3.15 **A line chart showing the variation over time of the unemployment rate in Canada. The reader should be aware of the fact that the Y-axis does not start at zero, which may give the impression that the variations are greater than they really are. An awareness of this can guard us against misinterpretations. (Source: Statistics Canada)**

the zero level of the quantity, or in drawing the Y-axis shorter than it should be. This procedure has the effect of giving the impression that the variations are bigger than they really are, but at the same time it allows us to see the variations in the graph in greater detail. When it is necessary to show a shorter Y-axis, this should be indicated by an interruption in the line representing the Y-axis. Figure 3.15 provides an example of a line chart. Here you can see that the Y-axis does not start at zero, giving the impression that the variations are much bigger than they really are. However, this is justified by the fact that it allows us to see the variation in unemployment rates in great detail, and by the fact that this is an increasingly standard practice, which means that readers should be aware of the resulting distortion and interpret what they see accordingly.

The General Shape of a Distribution

In addition to the measures explained above, we could describe the general shape of the distribution of a quantitative variable by looking at two of its features: *symmetry* and *kurtosis*.

Symmetry

The first characteristic to look at is *symmetry*. A distribution is said to be **symmetric** if the mean splits its histogram into two equal halves, which are mirror images of each other. A typical symmetric distribution is the *normal distribution*. It is a bell-shaped distribution that follows a very specific pattern, and occurs in a wide range of situations. It is represented by the curve of Figure 3.16. It will be studied later on.

In a symmetric distribution, the mean and the median are equal. If the distribution is also unimodal, then the mean, the median, and the mode are all equal. This is true of normal distributions.

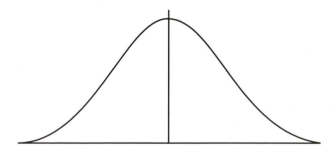

Figure 3.16   **An example of a symmetric distribution. This one is the normal distribution, which will be studied in Chapter 5**

If a distribution is symmetric and unimodal, the mean is a good representative of the center. However, it often happens that a distribution is not symmetric. We then say that it is **skewed**. That means that one side of the graph of the distribution is stretched more than the other. We say that it is **positively skewed** if it is stretched on the right side, and **negatively skewed** if it is stretched on the left side. Figure 3.17 illustrates the difference between symmetric distributions and skewed distributions. SPSS allows you to compute a statistic called **skewness**, which is a measure of how skewed a distribution is. A normal curve has a skewness of 0. If the skewness is larger than 1, the shape starts to look significantly different from that of a normal curve.

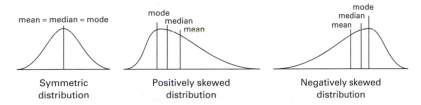

Figure 3.17   **Symmetric and skewed distributions**

How can we know that a distribution is skewed? The first indication is the histogram: the tail end of the histogram is longer on one side than on the other. We can also see that a distribution is skewed through its numerical features: the mean is different from the median. When the distribution is positively skewed, the mean is larger than the median, as it is pushed by the extreme values toward the longer tail. For negatively skewed distributions, the mean is smaller than the median. Therefore, a mean larger than the median tells us that the extreme values on the higher end of the distribution are much larger than the bulk of the data in the distribution, pulling the mean toward the positive side. This is illustrated by the numerical example given in the section on the median, where one extreme value (60) pulls the mean up but does not affect the median. Therefore, when a distribution is highly skewed, the median is usually a better representative of the center of the data than the mean.

Kurtosis

This is a measure of the degree of peakedness of the curve. It tells you whether the curve representing the distribution tends to be very peaked, with a high proportion of data entries clustered near the center, or rather flat, with data spread out over a wide range. A normal distribution has a kurtosis equal to 0. A positive value indicates that the data is clustered around the center, and that the curve is highly peaked. A negative value indicates that the data is spread out, and that the curve is flatter than a normal curve. Figure 3.18 shows three curves with zero, positive, and negative kurtosis respectively.

Methodological Issues

Although they seem to be simple, descriptive measures can be tricky to use. We would like to point out here some of the pitfalls and difficulties associated with their use.

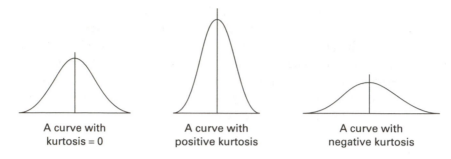

Figure 3.18   **Illustration of zero, positive and negative kurtosis**

The Definition of the Categories over which the Counting is Done

Suppose I say that the passing rate in a given class is 82%. In another college, a colleague tells me that his passing rate is 95%. Before concluding that his passing rate is much higher, I have to make sure that we are defining the passing rate in the same way. I may define the passing rate as the number of students who pass a course compared to those who were registered at the beginning of the semester. If he defines it the same way, we can make meaningful comparisons. But if he defines it as the number of students who pass the course compared to the number registered at the end of the semester, we cannot make a meaningful comparison. This is so because all the students who dropped out would not be taken into account in his calculation, whereas they would be taken into account in mine. A careful definition of the categories used to define a concept is therefore important. Such problems arise when we define the unemployment rate in various countries, or even wealth. The conclusion is that careful attention should be given to the way categories are defined when comparing the statistics that refer to different populations.

Outliers

Outliers are values that are unusually large or unusually small in a distribution. They have to be examined carefully to determine if they are the result of an error of measurement, or a typing error, or whether they actually represent an extreme case. For instance, the value 69 in the column of the variable **age** for college students could be a typing error, but it could also represent the interesting case of a retired person who decided to pursue a college program. Even if they represent an extreme case, it may be desirable to disregard extreme values in some of the statistical computations.

When producing a Box Plot diagram, SPSS excludes the outliers from the computation, and prints them above or below the box plot. An option allows users to have the case number printed next to the dot representing the outlier, so as to be able to identify the case and examine it more closely.

Summary

We have seen in this chapter the various measures used to summarize the data pertaining to a single variable as well as the various types of charts that could be used to illustrate the distribution. You should keep in mind one fundamental point: *the level of measurement used for the variable determines which measures and graphs are appropriate*. It does not make sense, for example, to compute the mean of the variable when the level of measurement is nominal, that is, when the variable is qualitative.

There are three types of univariate descriptive measures:

- measures of central tendency,
- measures of dispersion, and
- measures of position.

Measures of central tendency, also called measures of the center, tell us the values around which most of the data is found. They give us an order of magnitude of the data, allowing comparisons across populations and subgroups within a population. They include the mean, the median, and the mode. The mean should not be used when the variable is qualitative.

Measures of dispersion are an indication of how spread out the data is. They are mostly used for quantitative data. The most important ones are the range, the interquartile range, the variance, and the standard deviation.

Measures of position tell us how one particular data entry is situated in comparison to the others. The percentile rank is one such measure. Other measures include the quartiles and the deciles.

In addition to these measures, we have seen the weighted mean. When calculating it, the various entries are multiplied by a weight, which is a positive number between 0 and 1. All the weights add up to 1. The weighted mean is used when the numbers that are averaged have been calculated over populations of unequal size. For instance, if you have the birth rates in all Canadian provinces and you want to find the average birth rate for Canada as a whole, you must weight these numbers by the demographic importance of every province. The weighted mean is also used when you want to increase or decrease the relative importance of the numbers you are averaging, as is done when finding the average grade over exams that do not count for the same percentage in the final grade.

When categories are involved (either because the variable is qualitative, or when quantitative values have been grouped) we can find ratios, percentages, and proportions of the groups corresponding to the categories.

The general shape of a distribution is analyzed in terms of symmetry or skewness, and in terms of kurtosis (the degree to which the curve is peaked).

The comparison of the mean and the median is very useful. Recall the following:

> If the distribution is very skewed, the median is a better representative of the center of the data, as the extreme values tend to pull the mean towards one side of the curve. The median is not affected by extreme values.

> If the mean is larger than the median, the distribution is positively skewed. If the mean is smaller than the median, the distribution is negatively skewed.

As for the graphical representation of a distribution, recall again that the level of measurement of the variable determines what kind of chart is appropriate. Bar charts and pie charts are appropriate when the data is qualitative, or measured at the nominal or ordinal levels. Quantitative data (whether measured at the ordinal of numerical scale levels) could also be represented by bar charts or pie charts if the values have been grouped into a small number of categories.

The essential difference between pie charts and bar charts is that in the former, the emphasis is on the relative importance of each category as compared to the other categories, whereas in the latter, the emphasis is on the size of each category. However, there is no clear-cut distinction between the two, and if one is appropriate, the other is usually appropriate also, even if the emphasis is slightly different. The great advantage of bar charts is that it allows making comparisons between the distributions of subgroups, with the help of clustered bar charts.

Quantitative variables are better represented through histograms. A specific type of histogram is the population pyramid, which is a standard tool in demography.

Line charts are most suited to represent the variation of a quantity across time.

In all kinds of charts, truncating the Y-axis is sometimes done to zoom in on the variations of the variables and to represent them in a more detailed way. However, we should be aware of the fact that truncating the Y-axis may also convey a mistaken impression that the variations of the variable are more important than they are in reality.

Keywords

Univariate	Frequencies	Ratios
Bivariate	Cumulative frequencies	Proportions
Measures of central tendency	Valid percent	Bar graph
Measures of dispersion	Range	Clustered bar graph
Measures of position	Trimmed range	Pie chart
Mean	Interquartile range	Histogram
Trimmed mean	Deviation from the mean	Frequency polygon
Weighted mean	Standard deviation	Line chart
Median	Variation ratio	Box plot

Mode	Coefficient of variation	Five-number summary
Modal category	Quartiles	Symmetry
Majority	Deciles	Skewness
Plurality	Percentiles	Kurtosis
	Percentile rank	Outliers

Suggestions for Further Reading

Devore, Jay and Peck, Roxy (1997) *Statistics, the Exploration and Analysis of Data* (3rd Edn). Belmont, Albany: Duxbury Press.

Harnett, Donald H. and Murphy, James L. (1993) *Statistical Analysis for Business and Economics*. Don Mills, Ontario: Addison-Wesley Publishers.

Trudel, Robert and Antonius, Rachad (1991) *Méthodes quantitatives appliquées aux sciences humaines*, Montréal: CEC.

Wonnacott, Thomas H. and Wonnacott, Ronald J. (1977) *Introductory Statistics* (3rd edn). New York: John Wiley and Sons.

EXERCISES

3.1　Complete the following sentences:

(a) Three types of measures are useful to summarize a numerical distribution. They are

_____,

_____　and

_____.

(b) The most frequent value in a distribution is called _____.

(c) When the values of the distribution are grouped into classes, the mode is the _____ with the highest frequency.

(d) When there are two classes that are bigger than the ones immediately next to them, the distribution is called _____.

(e) If the modal class includes more than 50% of the population, we say that it constitutes the _____. Otherwise, we simply talk of a

_____ .

(f) The median falls _____ of the ordered list of entries. _____% of the data are less than or equal to the median, and _____ % are larger than or equal to it.

(g) The mean of a numerical distribution is equal to the _____ of all entries divided by _____.

(h) The mathematical measure used to find the mean when the entries do not have the same relative importance is called _____.

(i)   Quartiles separate the ordered list of entries into _____ equal parts. _____ % of the data are smaller than or equal to the first quartile, and _____ % of the data are larger than the first quartile or equal to it.

(j)   The numerical values that separate the ordered list of entries into 100 equal parts are called _____ . The _____ of a given entry is the hundredth part in which it falls.

(k)   When the mean, median, and mode coincide, and if the two halves of the curve representing a distribution are mirror images of each other, the distribution is said to be _____ .

If the mean is larger than the median, the curve is stretched to _____ and the distribution is said to be _____ _____ . If the mean is smaller than the median, the curve is stretched to _____ and the distribution is said to be _____ .

Numerical Exercises

3.2   Represent the following data values on a line, compute the mean, median, and mode (if the latter exists), and determine whether the distribution is symmetric or not.

(a)   1, 5, 6, 4, 2, 7, 11
(b)   1, 4, 5, 5, 5, 7, 8

3.3   Find the mean of the following distribution consisting of grouped data. Use the midpoint of every class to compute the mean.

Class i	Frequency f_i	Midpoint x_i	$f_i * x_i$
20 to less than 30	8		
30 to less than 40	21		
40 to less than 50	29		
50 to less than 60	18		
60 to less than 70	11		
Sum			
		Mean:	

3.4   (a)   For the following distribution, determine the value of N, and the various x_i. Write the formula for the mean, and replace the symbols by their values to compute the mean. Find also the median.

$$3, 6, 5, 8, 11, 2, 4, 9, 12, 8$$

(b)   For the following distribution, determine the value of N, and the various midpoints x_i and the frequencies f_i. Write the formula for the mean, and replace the symbols by their values to compute the mean. Find also the class in which the median falls.

Class	$0 \leq x < 5$	$5 \leq x < 10$	$10 \leq x < 15$	$15 \leq x < 20$
Frequency	7	12	22	13

3.5 The estimated price of adjacent houses in a street is given as follows.

$79,000	$83,500	$78,000	$90,000	$85,500
$87,000	$92,000	$81,000	$99,000	$384,000

Find the mean and the median of this distribution. Which of these two measures is more representative of the central tendency? Explain why.

3.6 For the following examples, determine the level of measurement of the variable (that is, the type of measuring scale used) and then determine the best measure to represent the central tendency of the data.

(a) The number of customers who entered a shop each day over a week is given by: 44, 60, 44, 50, 55, and 72.

(b) The customers that entered a given shop were distributed as follows:

Students: 129;
Businessmen and businesswomen: 14;
Unemployed: 15;
Employees: 167.

3.7 The following table gives the mean and the median of the grades obtained in a course for three classes of students.

Class	Mean grade	Median grade	Class size
A	69	59	33
B	74	70	25
C	80	75	22

(a) Find the overall average grade for the three classes combined (use the formula for weighted means).

(b) Is it possible to find the median for the three classes combined? Explain your answer.

(c)   Is it possible to determine how many students exactly passed the course in each class? (The passing grade is 60.) Is it possible to determine whether a majority of students passed the course in each class? Find these numbers whenever possible.

3.8   For the following distributions determine at sight, without the use of calculator, which of the proposed numbers are possible values for the mean.

Distribution		Proposed answer		
2, 4, 4, 6, 9, 11	a) 11	b) 3	c) 1.23	d) 6
24, 26, 32, 35, 36, 40, 87	a) 100.7	b) 15	c) 40	d) 38

3.9   Studies in the field of psychology have concluded that first-born children in a family are more likely to be great achievers in their studies than their younger brothers and sisters. A school psychologist wants to assess whether the grades and the IQ (intelligence quotient) differ markedly between students who are first-born in their family and other students. She compiles the following data:

First born		Others	
Grade	**IQ**	**Grade**	**IQ**
85	116	70	119
78	113	64	112
82	112	65	114
76	121	78	116
84	124	74	126
78	118	80	121
77	119	82	120
82	122	68	114
93	125	52	109
88	120	90	127

(a)   Compute the mean and the median of the grades and of the IQs for the two groups of students.

(b)   Is the difference between the two groups large enough in your opinion? (We will see later in the manual that there is a rigorous way of answering this question.)

(c)   Is the mode useful for this discussion? Explain why.

3.10   A study is done on the number of grammatical and spelling mistakes done by 2385 students in an English essay in their first semester of study at the college. The essays were two pages long and contained an average of 250 words each. The data is compiled in the following table.

Number of mistakes	Frequency
$0 \leq x < 5$	328
$5 \leq x < 10$	646
$10 \leq x < 15$	577
$15 \leq x < 20$	346
$20 \leq x < 25$	223
$25 \leq x < 30$	148
$30 \leq x < 35$	48
$35 \leq x < 40$	30
$40 \leq x < 45$	21
$45 \leq x < 50$	7
$50 \leq x < 55$	4
$55 \leq x < 60$	4
$60 \leq x \leq 80$	3
Total	2385

Answer the following questions

(a)   What are the appropriate measures of central tendency that could be used to summarize the data?

(b)   What is the most frequent number of mistakes that is found in one essay?

(c)   What is the median number of mistakes found in one paper?

(d)   On the average, how many mistakes does a student in the first session make per essay?

(e)   We want to come up with a statement of the form: 'In these essays, we find on the average one mistake every … words.' Is it possible to find the missing number? Explain.

(f)   Draw a histogram of the distribution of mistakes. How do the mean, median, and mode compare? What can you say about the general shape of the distribution?

(g)   Complete the following statement: 'The weakest 5% of students make … mistakes per essay or more.'

3.11   The student association of a college wants to examine the program of loans and scholarships given to its membership. It comes up with the following data. Take into account that in order to get a scholarship, you must first be eligible to get a loan, and only some of those who get a loan are eligible to get a scholarship.

Year	1995–1996	1996–1997	1997–1998
Number of applications received	2092	1992	1822
Number of applications rejected	184	253	296
Amount given in loans	$2,265,386	$2,708,660	$2,296,665
Amount given in scholarships	$2,571,973	$2,069,085	$1,848,650
Total amount of financial aid	$4,837,359	$4,777,745	$4,145,315
Number of full-time students	6092	6063	6171

(a) Which measures of central tendency can we use to summarize the data?

(b) For each year, compute the average amount of financial aid a student gets. Can you see a trend over those three years? (The average must be calculated over the total student population.)

(c) Among the students who get a loan, what is the average amount of loan a student gets? Can you see a trend over the three years? (Here the average is computed not over the total student population, but over the number of students who get a loan.)

(d) Does the data given allow you to find out the average amount a student in this college gets in the form of a scholarship? Does it allow you to compute the average value of a scholarship (that is, among those who do get a scholarship)? Are we missing some other data to be able to compute that the average amount of a scholarship?

(e) Determine which of the following sentences reflects the observed trends. If none does, try to formulate one that does reflect the observed trend.

  • The proportion of students who receive some financial assistance is on the rise, and the average amount given per student is also increasing.
  • The proportion of students who receive some financial assistance is on the rise, but the average amount given per student is decreasing.
  • The proportion of students who receive some financial assistance is in decline, but the average amount given per student is increasing.
  • The proportion of students who receive some financial assistance is in decline, and the average amount given per student is also declining.

3.12 The following table gives the distribution of employment positions within a small company, together with the average salary received by each group.

Position	Annual salary ($)	Frequency
CEO (Chief Executive Officer)	250,000	1
Management	90,000	5
Engineers	50,000	6
Foremen	30,000	10
Skilled workers	20,000	80
Unskilled workers and clerks	15,000	88
Support staff	12,000	10
Total:		200

(a) What is the population under study?
(b) Specify the two variables included in this table. For each, determine the type of variable and the level of measurement.
(c) What are the measures of central tendency that could be used for each of the variables? Calculate that measure for the data given in this table.
(d) Which of these measures are more representative of the data given in the table?

4 WRITING A DESCRIPTIVE SUMMARY

The purpose of this chapter is to explain how to proceed in order to write a good descriptive report, and how to analyze a frequency table beyond a first-level reading of the percentages, in order to identify the numerical features of the data and to highlight them.

After studying this chapter, the student should know:

- how to proceed when writing a descriptive report to summarize data;
- which measures and charts are appropriate, depending on the measurement level of the variable;
- how to summarize a set of variables that measure a given concept;
- how to analyze a frequency table in detail and identify its important features;
- the difference between a first-level description and an analytical description;
- the criteria for a good descriptive summary.

In Chapter 3, we have seen how to produce simple descriptive statistical measures, as well as simple tables and graphs. We have also seen that the statistical measures to be used depend on the level of measurement of the variable. Now, we would like to see how we can integrate all these elements and produce a synthetic report that describes certain features of a population. For the time being, we will restrict these explanations to univariate descriptions of variables. Later on, you will have to include bivariate descriptions, that is, descriptions of the statistical associations between variables, as well as confidence statements, that is, generalizations from the observed sample to the population as a whole, two statistical topics studied later on in this book. We will also learn how to report the result of a hypothesis testing.

How to Write a Descriptive Report

We will consider two types of report. Basic reports consist in a direct reading of the tables produced by SPSS, and a reformulation in direct, plain language of what the tables say, with accompanying charts as illustrations. There is very little interpretation in this case. A second level in sophistication consists in writing analytical reports: such reports would highlight the outstanding tendencies that can be seen in the data, and may include a greater degree of interpretation. We will now explore both kinds of reports.

Basic, Direct Reports

Suppose you want to describe the educational level of the individuals included in the **GSS93 subset** data file supplied with the SPSS package. This means that you would like to have some global description that tells you whether the people in your sample tend to have a high level of education or not (this is a description of the central tendency), and whether there is a big polarization, with some people having a lot of education and many others very little (this is a description of the dispersion).

The first thing to do is to see which variables concern education. You will find three such variables in the **GSS93 subset** data file. List them, and list the level of measurement of each.

In this data file, you will find that the three variables are:

- Highest year of schooling completed (scale),
- Highest degree obtained (ordinal, 5 categories), and
- Possession or not of a college degree (ordinal, 2 categories).

Determine what kind of descriptive measures you would use for each. Would you use a frequency table? For which of the variables? Which charts would be more appropriate?

Sometimes you will feel that you are not too sure which type of chart is appropriate. Get SPSS to produce several charts, examine them carefully to see which ones convey a better representation of the distribution of the variable, then select one of them, and paste it into your report.

One of the important pitfalls that you should avoid is to give a lot of tables or charts that are not very useful. You may want to be selective here: select the relevant information, and try to write it in a clear and concise way. For example, SPSS produces tables giving you the number of valid answers. You do not need to include the table itself. You could simply write in brackets ($n = 1500$) when describing the sample, to indicate that your sample contains 1500 individuals. Whenever you discuss or describe the results that relate to one of the variables, if you see that there are a lot of missing answers, add a phrase about the number of valid answers, such as (valid $n = \ldots$) and fill in the number of valid answers. Although the number of people in the sample is the same throughout the analysis of this data file ($n = 1500$), the number of valid answers varies a lot. This is why you have to specify how many *valid* answers you have to a particular question. You do not have to do that for every single question: you report the number of valid answers only when there is a lot of missing data, and the valid percentages differ by several points from the total percentages. It is advisable in this case to report the valid percentages. In some cases it may be relevant to report both the valid and total percentages.

What follows is a set of criteria that define a good descriptive report.

Criteria for a Good Report

THE GENERAL PRESENTATION

Make sure the text is clear, well organized, and concise. If the analysis is long, a cover page may be desirable. Make sure that all the relevant information is in it: a title, your name, the name of the course and the course number, the name of the instructor to which you are presenting it, and the date.

Some of this information, such as your name and the assignment number, could be written in the header of your document (refer to Lab 2 for explanations on the header). The tables and graphs must be printed with the correct identification: a title must be given to every table or graph. If you copy the tables from SPSS with the **Copy**... command (rather than the **Copy Object**... command), you can edit the table, and delete the rows or columns that are not useful or relevant. Also avoid grammatical mistakes: a spell check may be useful, but rely always on a careful reading of your report.

Include in your report a **description of the data file** you are using: its source, the year the survey was conducted, the kind of variables that are found in it, the institution under which it was conducted, etc.

DESCRIPTION OF THE VARIABLES UNDER STUDY

Make sure to include in your study all the variables that are relevant for your subject. If there are several variables that address a given topic, use them all to analyze this topic. For instance, 'education' can be measured in several ways. If there are several variables that deal with education, examine the distribution of each.

To describe a variable properly, you must select the appropriate measures. Do not compute the mean of a qualitative variable, because it is meaningless. You may want to use some of the recoded variables, or recode some variables yourself. Do not include a table of frequencies if the variable is quantitative. Such tables are usually quite long, and they are not useful to the reader. If the quantitative variable has been grouped into a small number of categories, a frequency table may be useful, in addition to the descriptive measures used for quantitative variables. Finally, formulate your conclusions in full, grammatically correct sentences that highlight the meaning of your numerical results. An example of a very concise description of the educational level of the people in our sample is given in Insert 4.1.

The appropriate measures to be used are summarized in Table 4.1.

Table 4.1 **Appropriate descriptive measures for the various levels of measurement**

Level of Measurement	Appropriate Statistical Measures	Appropriate Charts
Nominal (categories)	Frequencies, percentages, mode. Ratios, proportions and rates.	Bar charts, pie charts
Ordinal	Frequencies; mode; median. Cumulative frequencies. (If there are many categories, you may compute the mean and median, but the interpretation of the numerical results may be problematic.)	Bar charts; histograms
Numerical scale, ungrouped	Mean, median, mode, range, minimum, maximum standard deviation, interquartile range. (Frequency tables are not useful for this type of measure.)	Histograms, frequency polygons, box plots, time lines
Numerical scale, grouped	Frequency tables, mode. If there are a large number of groups: mean and standard deviation. The mean is usually the mean code of the categories. It can be used for comparative purposes if other samples are grouped in the same way, but it should not be mistaken for the mean of the variable itself. If grouped into a small number of categories, it should be treated like ordinal data.	Histograms, bar charts, pie chart. Box plots may be misleading if the number of categories is small.

Examples of Concise Descriptive Reports

What follows (Insert 4.1) is an example of a short descriptive report, which answers the question: Describe the educational level of the sample given in the file **GSS93 subset** that comes with the SPSS program.

INSERT 4.1 Descriptive report of the educational level of the sample

The data set used here is a subset of the General Social Survey conducted in the US in 1993 ($n = 1500$). There are three variables in this data set that address the issue of education: the highest year of schooling completed (scale), the highest degree obtained (ordinal, 5 categories) and the possession or not of a college degree (ordinal, 2 categories).

The *average* highest year of schooling completed is 13 years with a standard deviation close to 3 years. The graph below shows the distribution of this variable.

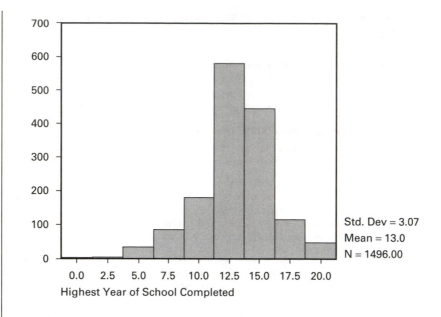

There is a slight difference between men and women in this sample. The average number of years of schooling for men is 13.2 years, whereas it is 12.9 years for women, a situation that can be represented by the following box plot.

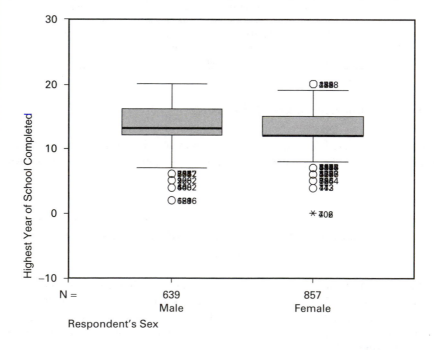

As for the highest degree obtained, the frequencies are the following:

Respondent's Highest Degree

	Frequency	Valid Percent	Cumulative Percent
Less than HS	279	18.6	18.6
High school	780	52.1	70.8
Junior college	90	6.0	76.8
Bachelor	234	15.6	92.4
Graduate	113	7.6	100.0
Total valid	**1496**	**100.0**	

We can see that 18.6% of the individuals did not complete high school, and that 52.1% of them stopped at high school. A total of 91.4% have completed high school or more.

The variable College Degree simply regroups these categories, lumping together those who do not have a College degree (76.8%) and those who do (23.2%). If we break this figure by sex, we find that 27.3% of the men of this sample have a College degree, while 20.1% of the women have one.

Another example is given in Insert 4.2.

INSERT 4.2   Example of a descriptive summary describing the family relationships for the individuals in the GSS93 subset data file

THE FAMILY RELATIONSHIPS OF RESPONDENTS IN THIS SAMPLE

1. The Data File

The data that will be analyzed comes from the General Social Survey conducted in 1993 in the US. This file contains 1500 cases and 67 variables. The variables contain some socio-economic data (marital status, working status, age, income, etc.), some variables that relate to the opinions of the respondents about social issues (such as gun control, abortion, or capital punishment) and about their political views (such as party affiliation), and some variables about their leisure and their musical taste.

2. The Variables Studied

We are going to examine closely the variables that deal directly with family relations. These are:

- **Marital status** (nominal, 5 categories). This variable has been recoded as **Married?** with two categories: Yes and No. The 'No' includes the categories: Never Married, Separated, Divorced and Widowed.
- **Age when first married** (scale)
- **Number of children** (scale). We will compare this variable with the variable Ideal number of children (scale).
- **Number of siblings** (scale)

To the extent that it deals with the relation between the parents and their children, the variable **Spanking children** (ordinal, with 4 levels) could also be included in the analysis, but it could be argued that this is not a family relationship, and that it belongs to the moral outlook of the respondents.

3. Marital Status

We get the following frequency table for the variable Marital Status:

Marital Status

	Frequency	Percent	Valid Percent	Cumulative Percent
Married	795	53.0	53.0	53.0
Widowed	165	11.0	11.0	64.0
Divorced	213	14.2	14.2	78.3
Separated	40	2.7	2.7	80.9
Never married	286	19.1	19.1	100.0
Total valid	**1499**	**99.9**	**100.0**	
NA	1	.1		
Total	**1500**	**100.0**		

We see that 53.0% of the individuals in this sample are married, 11.0% are widowed, 14.2% are divorced, and 2.7% are separated. Thus, those who have been married at some point add up to 80.9% while those who have never been married constitute 19.1% of the population. The variable Married? groups together in a single category all the individuals who are not currently married, and who add up to 47.0% of the sample.

4. Age When First Married

The average age when first married is 22.8 years of age. If we analyze this variable separately for men and women, we get the following table:

Descriptives: Age When First Married

Respondent's Sex		
Male	Mean	24.16
	Median	23.00
	Std. Deviation	4.867
	Minimum	16
	Maximum	50
Female	Mean	21.84
	Median	21.00
	Std. Deviation	4.929
	Minimum	13
	Maximum	58

We see from the table that men tend to get married later than women; the mean is 24.2 years for men, and 21.8 years for women. Half the men had already been married by the age of 23 years (that's the median). Half the women had been married by the age of 21 years. The youngest man to be married was 16, whereas the youngest woman was 13. The oldest man to be married for the first time was 50, while the oldest woman was 58. We also see that the standard deviation for men and women is almost the same.

The situation can be illustrated by the following box plot.

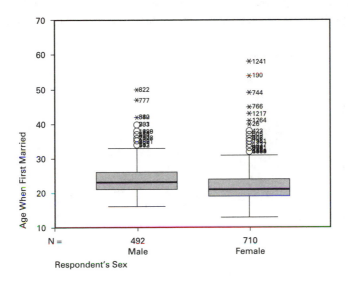

5. Number of Children

The individuals in this sample have on the average 1.85 children. If we examine the frequency table for that variable, we see that 27.7% of the people do not have children at all. Among those who have children, the most common situation is to have 2 children. Large families are not common, as in 93.3% of the cases individuals have 4 children or fewer. The situation can be represented by the following histogram.

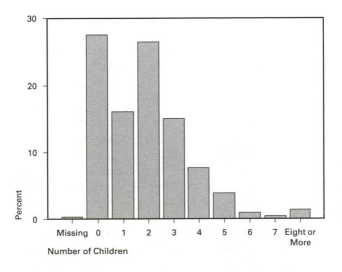

If we compare that situation with the Ideal number of children, we see that the mean for that variable is 2.76 children, but the comparison with the actual number of children is difficult to make, as there are 535 missing answers for that variable (we can assume that only those who had children were asked that question). It is better to examine the histogram of the ideal number of children. Here we see that the mode, or most desirable situation, is by far the situation with two children. Very few people think that one child is the ideal situation.

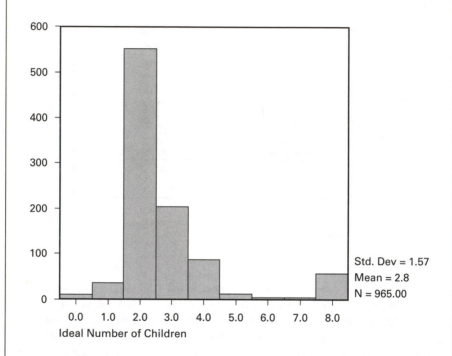

Std. Dev = 1.57
Mean = 2.8
N = 965.00

Ideal Number of Children

6. Spanking Children

We have answers for 66% of the respondents, and the rest of the answers are missing. Of those who answered, about three-quarters (73.3%) indicated they either agree or strongly agree with spanking children as a disciplinary measure, while the rest (26.7%) disagree or strongly disagree.

7. Number of Siblings

We see here that the average is 3.7 brothers and/or sisters. If we examine the cumulative frequencies, we see that 60.2% of the respondents come from families of 4 children or less (the respondent plus 3 brothers or sisters), the rest (almost 40%) coming from families with 5 children or more. Comparing that with the number of children people currently have, we see that in general, individuals come from families that are larger than the families they themselves establish, since the average number of children in this sample tends to be much smaller than the number of brothers or sisters respondents have.

Analytical Descriptive Reports

The examples shown above are quite direct, and consist essentially in reporting, almost as is, the information provided in the frequency tables. But a more analytical view would permit a richer reading of such tables. To illustrate what is meant by that we will go into a more detailed – and more analytical – reading of frequency tables.

EXAMPLES OF HOW TO ANALYZE A FREQUENCY TABLE

To make our point clear, we are going to analyze four cases of the same situation, represented by the tables below. They all deal with the frequencies of the variable Political Party Affiliation, taken from the **GSS93 subset** file. The first table is the one that we get from the actual data in this file. The other three have been modified to illustrate how the analysis can highlight the distribution pattern.

Table 4.2 **Political Party Affiliation A**

	Frequency	**Percent**	**Valid Percent**
Strong Democrat	213	14.2	14.3
Not Str Democrat	298	19.9	20.0
Ind, Near Democrat	180	12.0	12.1
Independent	187	12.5	12.5
Ind, Near Republican	148	9.9	9.9
Not Str Republican	280	18.7	18.8
Strong Republican	168	11.2	11.3
Other Party	17	1.1	1.1
Total valid	1491	99.4	100.0
NA	9	.6	
Total	1500	100.0	

Case A   Analysis of Case A (Table 4.2). We see from the table that those who are affiliated with the Democrats (strongly or not strongly) add up to 34.3%, or slightly more than a third. Those who are affiliated with the Republicans add up to 30.1%, or slightly less than a third. The independents add up to 34.5, again a little more than a third. It is interesting to note that the population is almost evenly divided into three groups, and that those who affiliate to neither party are as numerous (or a little more numerous) than those who affiliate with either of the two main parties. We can also notice that, within each of the two main parties, those who do not have a strong affiliation with the party are more numerous than those who have a strong affiliation (for the Republicans: 280:168, or about 7:4, and for the Democrats, 298:213, or about 3:2). The bar chart shown in Figure 4.1 illustrates this situation.

Case B   Analysis of Case B (Table 4.3). We see from the table that those who affiliate with the Democrats add up to 42.1%. Those who are affiliated with the Republicans add up to 39.1%, or slightly less than the Democrats. The independents add up only to 17.6%, indicating that there is a strong polarization between the two

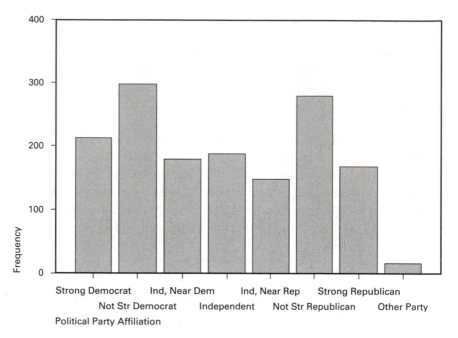

Figure 4.1 **Political Party Affiliation**

Table 4.3 **Political Party Affiliation B**

	Frequency	Percent	Valid Percent
Strong Democrat	272	18.2	18.2
Not Str Democrat	356	23.7	23.9
Ind, Near Democrat	122	8.1	8.2
Independent	57	3.8	3.8
Ind, Near Republican	84	5.6	5.6
Not Str Republican	351	23.4	23.5
Strong Republican	232	15.5	15.6
Other Party	17	1.1	1.1
Total valid	1491	99.4	100.0
NA	9	.6	
Total	1500	100.0	

parties, with less than 1 person out of 5 not affiliated to one of these two parties. We can also notice that, within a party, those who are not strongly affiliated with the party are more numerous than those who are (for the Republicans 23.4% vs. 15.5%, or a ratio of about 3:2, and for the Democrats 23.7% vs. 18.1%, or a ratio of about 4:3). The bar chart in Figure 4.2 illustrates this situation, and the polarization between the two parties is clearly visible.

Case C   Analysis of case C (Table 4.4). We see from the table that those who are affiliated with the Democrats add up to 35.6%, or slightly more than a third. Those

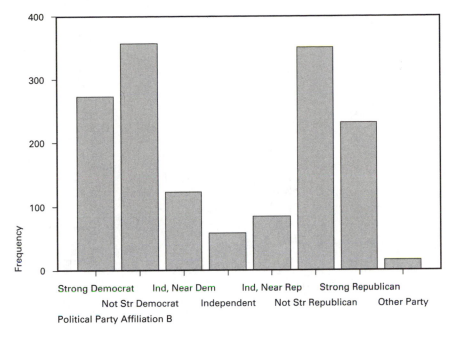

Figure 4.2 **Political Party Affiliation B**

Table 4.4 **Political Party Affiliation C**

	Frequency	**Percent**	**Valid Percent**
Strong Democrat	292	19.5	19.6
Not Str Democrat	236	15.7	15.8
Ind, Near Democrat	188	12.5	12.6
Independent	93	6.2	6.2
Ind, Near Republican	165	11.0	11.1
Not Str Republican	233	15.5	15.6
Strong Republican	267	17.8	17.9
Other Party	17	1.1	1.1
Total valid	1491	99.4	100.0
NA	9	.6	
Total	1500	100.0	

who are affiliated with the Republicans add up to 33.5%, or about a third. The independents add up to 29.9%. Thus, the population is almost evenly split between the three groups, with the Democrats only slightly ahead of the Republicans. Notice that, within each party, those who are strongly affiliated with the party are more numerous than those who are not (a ratio of 4:3 for the Democrats, and a ratio of 6:5 for the Republicans). This is illustrated in Figure 4.3.

Case D   Analysis of case D (Table 4.5). We see from the table that this is a situation of weak polarization between the Republicans and the Democrats. The Democrats attract 42.8% of the population, while the Republicans only get 30% of the

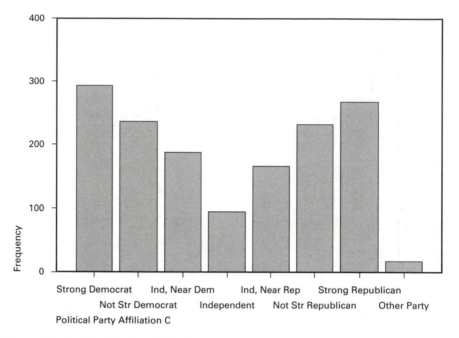

Figure 4.3   **Political Party Affiliation C**

support, almost 13 points behind the Democrats. The independents add up to 26.0% of the population. Notice that, within each party, those who are strongly affiliated with the party are the majority, with a ratio of about 4:3 for the Democrats and about 5:4 for the Republicans, a situation illustrated by Figure 4.4.

Table 4.5   **Political Party Affiliation D**

	Frequency	**Percent**	**Valid Percent**
Strong Democrat	356	23.7	23.9
Not Str Democrat	282	18.8	18.9
Ind, Near Democrat	188	12.5	12.6
Independent	116	7.7	7.8
Ind, Near Republican	84	5.6	5.6
Not Str Republican	202	13.5	13.5
Strong Republican	246	16.4	16.5
Other Party	17	1.1	1.1
Total valid	1491	99.4	100.0
NA	9	.6	
Total	1500	100.0	

As we have seen, the short descriptive paragraphs that follow each table do not simply report the frequencies. We have tried to highlight the specific features of each situation by answering the following questions: Is there a polarization

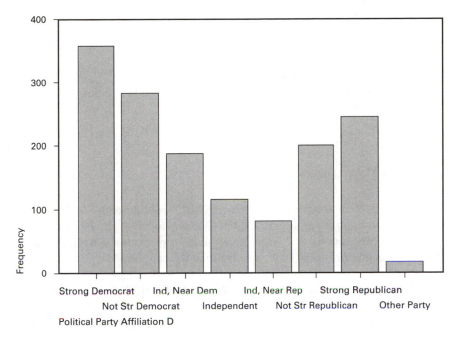

Figure 4.4   **Political Party Affiliation D**

between the two parties? Is one of them clearly more popular than the other? Is there a large proportion of independents? How is the level of mobilization *within* each party? We answered that last question by providing the ratio of those who feel a strong affiliation to the party compared to those who do not feel a strong affiliation.

A descriptive report that does that systematically is more analytical than one where the percentages are flatly reported as is. Insert 4.3 illustrates such a report.

**INSERT 4.3   Description of the Voting Behaviour
and of the Political Tendencies of a
Sample of US Residents**

The data summarized here come from a (non-representative) sample of 1500 individuals, which is a subset of the General Social Survey conducted in the US in 1993.

Four variables deal with our topic: Voting in 1992 Election, Political Party Affiliation, Think of self as Liberal or Conservative, and Political outlook. All four variables are measured at the nominal level. An examination of the frequency tables shows that the last variable is a recode of the third one, as explained below.

The following features can be observed:

- a low voter turnout;
- a polarization between the two main parties;
- a low level of mobilization within each of the two main parties;
- a large proportion of uncommitted voters.

Here are the details:

A LOW VOTER TURNOUT

Slightly more than two-thirds of the voters (68.2%) did make use of their right to vote, with a large group consisting of close to 30% (28.2% precisely) abstaining. Men and women voted in very similar proportions, with a slightly higher percentage of men (70.2%) voting than women (68.4%).

A POLARIZATION BETWEEN THE TWO MAIN PARTIES

The political affiliation is polarized between the Democrats and the Republicans, as shown in the chart below.

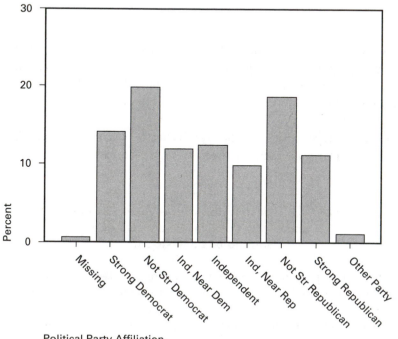

Political Party Affiliation

People who are affiliated with the Democrats or are close to them account for 46.4% of the sample, whereas those who affiliate with the Republicans or are close to them account for 40.0% of the sample, as shown in the table below.

Political Tendency	Valid Percent
Democrat or close to Democrat	46.4
Independent	12.5
Republican or close to Republican	40.0
Other Party	1.1
Total (*n* = 1491)	**100.0**

A LOW LEVEL OF MOBILIZATION WITHIN THE TWO MAIN PARTIES

Within each party, the level of mobilization is low, with many people feeling not strongly about their respective party. There are more people who feel not strongly Democrats (20.0% of the valid answers) than there are who feel strongly Democrats (14.3%), and the same goes for the Republicans, with 18.7% not strongly affiliated with the Party as opposed to 11.2% strongly affiliated with it, as shown in the table below.

Political Party Affiliation

Party Affiliation	Frequency	Percent	Valid Percent
Strong Democrat	213	14.2	14.3
Not Str Democrat	298	19.9	20.0
Ind, Near Democrat	180	12.0	12.1
Independent	187	12.5	12.5
Ind, Near Republican	148	9.9	9.9
Not Str Republican	280	18.7	18.8
Strong Republican	168	11.2	11.3
Other Party	17	1.1	1.1
Total Valid	**1491**	**99.4**	**100.0**
No Answer	9	.6	
Total	**1500**	**100.0**	

A LARGE GROUP OF INDEPENDENTS

If we only count those who feel strongly or not strongly affiliated with the Democrats, we get 34.3% for the Democrats, 30.1% for the Republicans, leaving 34.5% independent, or slightly more than one-third of the sample, which is a rather large group. This non-commitment reflects itself in the variable Think of Self as Liberal or Conservative, where more than a third of the respondents (36% of the valid answers) declare themselves moderate, the others being split almost evenly between the Liberals and Conservatives.

These comments summarize the main features of the variables under study.

Finally, we will see how a statistical agency such as Statistics Canada writes a descriptive report. Insert 4.4 is a report about marriages, which was published on October 28, 1999.

INSERT 4.4

The Daily. Thursday, October 28, 1999

MARRIAGES

1997

The number of marriage ceremonies conducted in 1997 fell to 153,306, down 2% from 1996 and 24% below the 1972 peak of 200,470.

Newfoundland, Saskatchewan and the Northwest Territories were the only regions where marriage ceremonies increased from 1996 to 1997 and these increases were slight (under 2%). All other provinces and territories reported declines, ranging from under 1% in Alberta and Quebec to 6% in New Brunswick and 15% in Yukon.

While the average age of brides was 30.9 years in 1997 – up from 28.4 years 10 years earlier – over half (56%) of the marriage ceremonies conducted in 1997 involved a bride aged 20 to 29. The average age of first-time brides was 27.4 years. Previously divorced brides averaged 39.8 years and previously widowed brides, 55.5 years.

The average age of grooms in 1997 was 33.5 years – an increase from 31.1 years 10 years earlier – with over half (52%) of the marriage ceremonies conducted in 1997 involving a groom aged 25 to 34. The average age of first-time grooms was 29.5 years. Previously divorced or widowed grooms were, respectively, 43.4-years-old and 62.1-years-old on average.

Three out of four brides and the same proportion of grooms were married for the first time in 1997. Most of the remaining brides (21%) and grooms (22%) had been previously divorced and the rest had been widowed (3% of brides and grooms).

In two out of three marriages in 1997, both spouses were marrying for the first time. A first-time partner married a divorced partner 18% of the time and both partners had been divorced in 12% of the marriages. Very few marriages in 1997 involved a widowed partner (under 5%).

Two-thirds (67%) of Canadian marriage ceremonies occurred in the warm weather months of May to September. August was the most popular month, accounting for 20% of all 1997 marriages.

The majority (76%) of 1997 marriage ceremonies were conducted by a member of the clergy; the remainder were solemnized by marriage commissioners, judges, justices of the peace and clerks of the court. There were striking differences among the provinces and territories in the type of officiant involved. In Ontario and the Atlantic Provinces, the majority of marriage ceremonies were conducted by clergy – 94% in Ontario and 80% to 86% in the Atlantic Provinces. In contrast, the far western region saw the majority of marriage ceremonies performed by officiants with no religious designation; only 29% of marriage ceremonies in Yukon and 44% in British Columbia were conducted by a member of the clergy.

When both spouses were marrying for the first time, the proportion of ceremonies conducted by members of the clergy climbed to 82%. In comparison, the proportion was down to 58% when both spouses had been previously divorced.

Marriages

	1996	1997	1996 to 1997
	Number		**% change**
Canada	156,691	153,306	−2.2
Newfoundland	3,194	3,227	1.0
Prince Edward Island	924	876	−5.2
Nova Scotia	5,392	5,177	−4.0
New Brunswick	4,366	4,089	−6.3
Quebec	23,968	23,958	−0.0
Ontario	66,208	64,535	−2.5
Manitoba	6,448	6,261	−2.9
Saskatchewan	5,671	5,707	0.6
Alberta	17,283	17,254	−0.2
British Columbia	22,834	21,845	−4.3
Yukon	197	167	−15.2
Northwest Territories	206	210	1.9

Summary

We have seen in this chapter that there are two levels of descriptive reports. The first level is a simple, direct reporting of the statistical results that are produced in SPSS. It is important to recall at this point that the level of measurement of the data determines the measures that are used to summarize it and the charts that are appropriate (refer to Table 4.1). The second level is more analytical. The specific features of the distribution, such as its symmetry or skewness, the polarization between groups, etc., must be identified and highlighted. These specific features can be proposed as the main conclusion of the descriptive report, which are then backed up by the tables and charts, as shown in Insert 4.3.

The descriptive reports we have learned how to write will be enriched, as we progress through the book, with the analysis of associations between variables, with inferences (confidence statements) and with the results of hypothesis testing.

Keywords

Basic descriptive report   Appropriate measures
Analytical descriptive report   Appropriate charts
Polarization   Specific features of a distribution
Ratio between groups

Suggestions for Further Reading

Howe, Renate and Ros Lewis (1993) *A Student Guide to Research in Social Science*.
   New York: Cambridge University Press.
Hult, Christine A. (1996) *Researching and Writing in the Social Sciences*. Boston: Allyn and
   Bacon.

EXERCISES

4.1   Write a short descriptive report to summarize the way people relate to leisure
      in the **GSS93 subset** file.
4.2   Write a short descriptive report to summarize the musical taste of respondents
      in the **GSS93 subset** file.
4.3. Write a short descriptive report to summarize the moral views of respondents
      in the **GSS93 subset** file.

5 NORMAL DISTRIBUTIONS

The purpose of this chapter is to explain the properties of the normal distributions, and to show how the percentage of data falling between any two values can be computed.

After studying this chapter, the student should know:

- what are normal distributions and when they mostly occur;
- the three properties that characterize them;
- how to read the table of areas under a normal curve;
- how to convert x-values into z-values and vice versa;
- how to compute the percentage of data between any two values in a normal distribution.

One type of distribution, called the normal *distribution*, occurs very often in social and natural phenomena. These distributions are symmetric, and they have the shape of the cross-section of a bell. Figure 5.1 shows a continuous variable, which is normally distributed, and Figure 5.2 shows a histogram, which approaches the shape of a normal distribution. The mathematical properties of normal distributions may sometimes be used to give approximate descriptions of such histograms.

Normal distributions can be described by the descriptive measures that we have seen so far. They are characterized by the following properties:

1. They are symmetric and unimodal (that is, they have a single mode), which means that the two halves of the distribution are mirror images of each other and that their mean, mode, and median are identical.
2. The graph that represents them is a bell-shaped curve.
3. The distribution can be completely described if we know that it is normal, and if we know its mean and standard deviation. For this reason, normal distributions are denoted by the symbols $N(\mu, \sigma)$. The N tells us we are talking about a normal distribution, the μ is the mean of the distribution and the σ is its standard deviation.

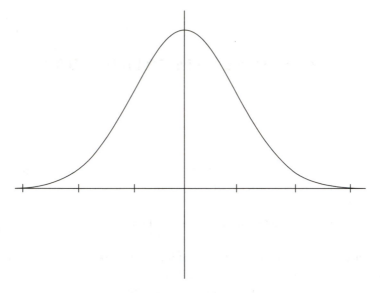

Figure 5.1

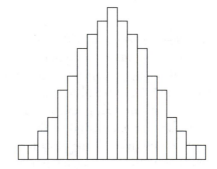

Figure 5.2

Properties of Normal Distributions

Normal distributions often occur when a quantitative variable is distributed at random. For instance, if we choose a random sample of, say, 3000 men, and we draw the distribution of their heights, we are likely to find the pattern of a normal distribution shown above.

All normal distributions share a common property: the number of data entries that fall between any two values depends on one thing only: the number of times you can fit a standard deviation between the mean and each of these values.

Look for instance at N(72, 4), and suppose it represents the distribution of weights of 3000 men, measured in kilograms. This means that the average weight is 72 kg, and the standard deviation is 4 kg. The value 76 is exactly one standard deviation

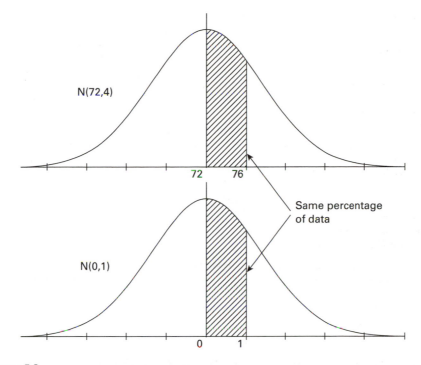

Figure 5.3

(4 kg) away from the mean (72 kg). Property 3 above means that the proportion of men whose weight is somewhere between 72 kg and 76 kg is exactly the same as the proportion of data falling between 0 and 1 in N(0, 1). This is illustrated by Figure 5.3.

Since all normal distributions follow the same pattern, we will always use **N(0, 1)** as the basic model for talking about normal distributions, and we will call it the **standard normal distribution**.

In order to distinguish individual values taken in a general normal distribution N(μ, σ) from those taken in the standard normal distribution N(0, 1), we will use different symbols to refer to them. The values on the X-axis of N(μ, σ) will be called *x-values*. The values on the X-axis of N(0, 1) will be called *z-values*, or **z-scores**.

The preceding discussion means that we can establish the following similarities between N(72, 4) and N(0, 1), shown in Table 5.1 on the following page (the student should complete the missing words or numbers).

If you look carefully at what you have been doing in Table 5.1, you will notice that to convert an x-value into a z-score, you look at how many times you can fit the standard deviation between the x-value and the mean. In other words, the z-score can be calculated as:

$$z = \frac{x - \mu}{\sigma}$$

We can also compute the x-value if we know the z-score: $x = \mu + z\sigma$

Table 5.1

In N(72, 4)		**In N(0, 1)**
(x-values)		(Corresponding z-scores)
The proportion of data falling between 72 and 80 in N(72, 4) (we can fit 2 standard deviations between 72 and 80)	is identical to	the proportion of data falling between 0 and 2 in N(0, 1)
The proportion of data falling between 72 and 78 in N(72, 4) (we can fit 1.5 standard deviation between 72 and 78)	is identical to	the proportion of data falling between 0 and 1.5 in N(0, 1)
The proportion of data falling between 72 and 77 in N(72, 4) (how many standard deviations can you fit between 72 and 77?)	is identical to	_____ _____
_____ _____	is identical to	the proportion of data falling between −1 and 0 in N(0, 1)
The proportion of data falling between 67 and 72 in N(72, 4)	is identical to	_____ _____

Using the Table of Areas Under the Normal Curve

We have mentioned that the proportion of data between two values in N(μ, σ) is the same as the proportion of data between the corresponding values in N(0, 1), but we have not said what this proportion was, numerically. Is it possible to compute such a proportion? The answer is Yes. This is done with the help of a table that tells us how many of the data entries are larger than any given z-value. Since all normal distributions follow the same pattern, one table is enough: the table for the standard normal distribution N(0, 1). This table is given in Appendix 1. Here is how we read the table. The table below (Table 5.2) is extracted from the full table given in the appendix:

Table 5.2

	0.00	**0.01**	**0.02**	**...**	**.....**	**0.08**	**....**
0.0	.5000	.4960	.4920	...		.4641	
0.1	.4602	.4562	.4522	...		.4286	
...							
1.4	.0808	.0793	.0778	...		.0694	
...							

The body of the table gives us the proportion of data that is larger than a given positive z-score. To look for the number corresponding to the z-score 1.48, you look at the number that sits in the intersection of the line 1.4 and the column 0.08. The table gives us a value of **0.0694**.

This means that in the normal distribution N(0,1), which has a mean equal to 0 and a standard deviation equal to 1, the proportion of data entries that are larger than 1.48 is equal to 0.0694. This situation is illustrated by Figure 5.4. We can formulate the proportion as a percentage instead: 6.94%, or close to 7% of the data is larger than 1.48.

We can see from the table in Appendix 1 that:

2.5% of the data is larger than $z = 1.96$;
0.5% of the data is larger than $z = 2.58$. (Figure 5.5)

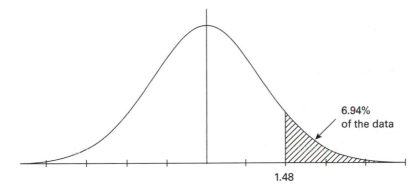

Figure 5.4

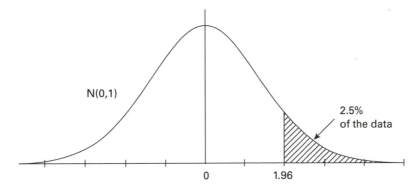

Figure 5.5

If we look at both halves of the diagram, these proportions must be multiplied by two and we can say that in the standard normal distribution N(0, 1):

5% of the data is farther than 1.96 units on each side;
1% of the data is farther than 2.58 units on each side. (Figure 5.6)

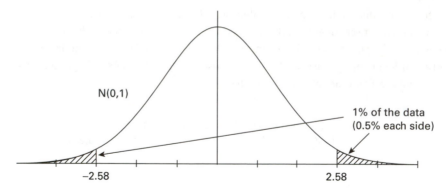

Figure 5.6

We can also say that in *any* normal distribution N(μ, σ):

> 5% of the data is farther than 1.96 standard deviations on each side of the mean;
> 1% of the data is farther than 2.58 standard deviations on each side of the mean. (Figure 5.7)

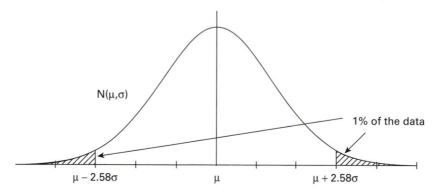

Figure 5.7

The same conclusion can be formulated in a different way, by looking at the data that falls *within* the z-values that are mentioned. We can thus conclude that:

> 95% of the data falls within 1.96 (or approximately 2) standard deviations each side of the mean;
> 99% of the data falls within 2.58 (or approximately 2.5) standard deviations each side of the mean.

This can be represented by the diagram in Figure 5.8.

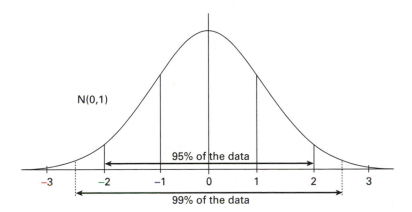

N(0,1)

95% of the data

-3   -2   -1   0   1   2   3

99% of the data

Figure 5.8

A large class of phenomena follow the pattern of a normal distribution. The distribution of a quantitative variable that is due only to randomness is generally normal. For instance, if we draw the histogram of the heights of a large group of men who have been selected at random from a homogeneous population, the shape of the histogram will be very close to that of a normal curve. If we compile the weights at birth of a thousand babies born naturally in a given hospital, the histogram representing the weights will have a shape very close to that of a normal curve.

The calculations done with the normal curve turn out to be very important for statistical inference, which will be discussed later on in this book (Chapters 9 and 10). The reason is that if we compute all possible samples of a certain size in a population, the distribution of their means (called the *sampling distribution of the mean*) has the pattern of a normal distribution. We will therefore be able to say: the means of 99% of the samples fall within certain values, and on this basis we will make a guess about the mean of the whole population. This will be explained further in Chapter 9.

Numerical Examples

The use of the table of areas under the normal curve, given in Appendix 1, allows us to solve the following exercises.

- Find the percentage of data in N(0, 1) that falls between $z = 0$ and $z = 1.32$.
  Solution: (Figure 5.9) Look in the table for the value corresponding to $z = 1.32$. You get 0.0934, which is the proportion of data that is larger than 1.32. Since the curve is symmetric, half of the data falls on the right-hand side of $z = 0$. Therefore:

  The proportion between 0 and 1.32 is = 0.5 − 0.0934 = 0.4066.

You can formulate the answer as a percentage:

  40.66% of the data falls between $z = 0$ and $z = 1.32$.

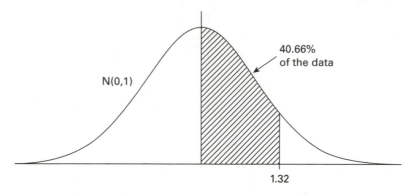

Figure 5.9

- In a provincial exam, it turns out that the average grade is 77 out of 100, with a standard deviation equal to 4 marks. Assuming that this is a normal distribution, find the percentage of students who got a grade of 85 or better.
  Solution: (Figure 5.10) The distribution we have is N(77, 4), and $x = 85$.
  We convert the x to a z-value: $z = (x - 77)/4 = (85 - 77)/4 = 2$.
  The table gives us: 0.0228, or 2.28%.
  Therefore, we can say that 2.28% of the students got a grade of 85 or better.

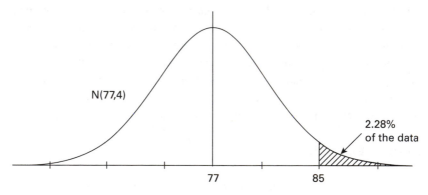

Figure 5.10

- Find the percentage of data in N(0, 1) that falls between $z = 0.5$ and $z = 1.2$.
  Solution: Draw a diagram representing the situation (Figure 5.11).
  We see that the required proportion is the difference between the two values given in the table, corresponding to $z = 0.5$ and $z = 1.2$ (find these values in the table).
  Therefore, the required proportion is

$$0.3085 - 0.1151 = 0.1934.$$

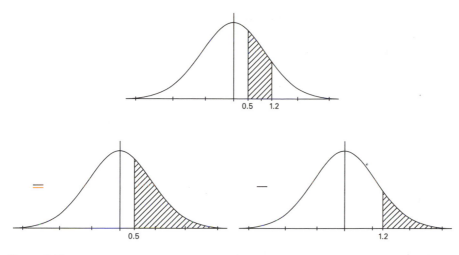

Figure 5.11

Therefore, we can say that 19.34% of the data falls between $z = 0.5$ and $z = 1.2$.

- You got 86 out of 100 in an exam, and you are told that the results for your whole school are normally distributed, with a mean of 80 out of 100, and a standard deviation of 4 marks. Would you say that you are in the top 5% of your school for this course? Or in the top 10%? Or in the top 1%?
  Solution: Convert your grade into a z-score, and check in the table. The corresponding z-score is 1.5, and the corresponding reading in the table is 0.0668, which means that the percentage of data larger or equal to this value is 6.68%. We conclude that this grade qualifies you for the top 10% of all grades, but not for the top 5%.

The pencil-and-paper exercises given in the exercise section will allow you to practice these computations.

Summary

Normal distributions are a very important and common pattern that occurs in a wide range of distributions of quantitative variables. All normal distributions follow basically the same pattern, symbolized by N(0, 1), a normal curve with mean equal to 0 and standard deviation equal to 1. The values in N(0, 1) are referred to as z-scores, and the table at the end of this chapter gives the exact proportion of data entries that are larger than any z-score.

Any value of a normally distributed variable can be converted into a corresponding z-score, by subtracting from it the mean, then dividing by the standard deviation.

This operation allows us to make use of the table to determine the proportion (and therefore the percentage) of data entries that fall between any two values, for any normally distributed variable.

Keywords

> Normal distribution
> Normal curve
> Standard normal distribution
> z-score
> Area under the normal curve

Suggestions for Further Reading

Devore, Jay and Peck, Roxy (1997) *Statistics, the Exploration and Analysis of Data* (3rd edn). Belmont, Albany: Duxbury Press.
Trudel, Robert and Antonius, Rachad (1991) *Méthodes quantitatives appliquées aux sciences humaines*. Montréal: CEC.
Wonnacott, Thomas H. and Wonnacott, Ronald J. (1997) *Introductory Statistics* (3rd edn). New York: John Wiley and Sons.

EXERCISES

For each of the following questions, draw a diagram of the normal curve and shade the area under consideration.

5.1   Given a standard normal distribution N(0, 1), find:
   (a)   the percentage of scores that are larger than $z = 1.42$;
   (b)   the percentage of scores that are less than $z = -0.6$;
   (c)   the percentage of scores that are less than $z = 1.42$;
   (d)   the percentage of scores that are between $z = -1.5$ and $z = 1.11$;
   (e)   a z-value such that 15% of scores are larger than it;
   (f)   a z-value such that 80% of the scores fall between $-z$ and $+z$.

5.2   Given a normal distribution N(230, 12), find:
   (a)   the percentage of scores that are larger than $x = 240$;
   (b)   the percentage of scores that are less than $x = 215$;
   (c)   the percentage of scores that are less than $x = 245$;
   (d)   the percentage of scores that are between $x = 212$ and $x = 242$;
   (e)   an x-value such that 15% of scores are larger than it;
   (f)   two x-values x_1 and x_2, equally distant from the mean, such that 20% of the scores fall *outside* of the region falling between x_1 and x_2.

6

SAMPLING DESIGNS

The purpose of this chapter is to explain the various types of samples, and to show how a simple random sample is constituted with the help of a table of random numbers. Errors due to sampling are also briefly explained.

After studying this chapter, the student should know:

- the importance of having a representative sample when the aim is to generalize;
- what characterizes the two broad categories of samples (probabilistic or non-probabilistic);
- the various types of samples in each category;
- how to select manually a simple random sample with the table of random numbers;
- the principles behind the selection of a quota sample;
- how to recognize which type a sampling procedure belongs to;
- how the total error in measurement is broken down into simpler types of errors.

We mentioned earlier (in Chapter 2) that data collection is a central component in research. If the data is not collected in a rigorous and systematic way, it cannot provide a basis for scientific knowledge.

Recall the distinction between the terms *statistic* and *parameter*.
A **statistic** (notice the singular) is any numerical measure computed or predicted on a sample.
A **parameter** is any measure computed or estimated on the whole population. Thus, when we compute the mean of a sample, we have computed a statistic, and we usually estimate the corresponding parameter, that is, the mean of the whole population.

When the purpose of the research is to draw inferences about a whole population on the basis of what has been observed on a sample, data collection becomes even more important. Recall that after we have collected the data that relates to a sample, we want to analyze it and then *generalize* it to the whole population. This is the ultimate goal of working with a sample: to generalize our findings to the whole population. A measure obtained from a sample is called a **statistic**, and the corresponding measure for the whole population is called the **parameter**. The parameter is estimated to be equal to the measured statistic, plus or minus a margin of error, which depends on the risk we are willing to take of being wrong: if we are not willing to take a big risk of error, we must allow for a large margin of error (this question will be discussed in more depth in Chapter 9, on estimation).

If the method used to select the sample tends to produce samples that are really *representative* of the whole population, the generalization will be made on a sound basis,

> The **sampling design** is a detailed plan for coming up with a sample, which specifies the type of sample used, the list of units from which the sample is going to be selected, the number of units needed, and the precise method of selecting them.

and we will be able to generalize our results with a high level of confidence that they are correct. But if the method used to select the sample is biased, that is, if it tends to produce samples that differ significantly from the population, our results will be rather shaky and not reliable. Hence the importance given to the study of how the sample is constituted and how the individual units in it have been selected. This is what we mean by *sampling design*. The sampling design refers to the way the sample is constituted and to the method used to select which individuals are in it. The sampling design is a detailed plan for coming up with a sample, which specifies the type of sample used (explained below), the list of units from which the sample is going to be selected (called the *sampling frame*), the number of units needed, and the precise method of selecting the units. The sampling design is thus a fundamental question in all types of quantitative research.

There are three principal types of research design that produce quantitative data: survey research, experimental research, and secondary data analysis, which we classified as a type of archival research. There are many more types of research design, such as direct observation, participant observation, life history analysis, content analysis, ethnographic research, etc. These other types may involve quantitative data, but their focus is rather qualitative. We will focus therefore on the three types that are more specifically quantitative. Archival research is often qualitative, but it is quantitative when it focuses on the analysis of secondary quantitative data, such as the analysis of Census data collected by most national statistical agencies in the various countries.

Types of Samples

There are two types of sampling design: those that are based on probability and those not based on probability. These notions will become clear by the end of this section. A temporary definition is the following: **in a probability sample, each unit has a known probability (that is, likelihood) of being selected, and the selection is based on a random choice of the units**. The non-probability samples are not based on a random choice and the consequence is that results based on non-random samples tend to be biased.

We will briefly explain the following types of samples.

- **Probabilistic samples:**

  - Simple random samples
  - Systematic samples
  - Cluster samples
  - Stratified random samples:

    - Proportional
    - Non-proportional

- **Non-probabilistic samples:**

  - Quota samples
  - Convenience samples
  - Judgment samples
  - Samples of volunteers

Probabilistic Samples

SIMPLE RANDOM SAMPLES

The simplest type of probability sample is the *simple random sample* or *SRS* in short. *A **simple random sample** is a sample chosen from a population by a random procedure, in such a way as to give each unit of the population exactly the same chances (probability) of being selected.* For instance, if you assign an identity number to each of your units, you write the numbers on identical spherical balls that are put in a bag and thoroughly mixed, and then you blindly choose a fixed number of balls, you are producing a simple random sample. Such a procedure can be performed with repetition (by putting a ball that has been selected back into the bag) or without repetition (by excluding a chosen ball from subsequent selections). If you choose a sample of n units without repetition out of a population of N units, each unit in the population has exactly n chances of being selected out of N, that is n/N. This probability of being selected is the same for all units.

There is a mathematical method that replicates the procedure used for choosing a random sample that we have just explained. It is based on the table of random numbers. It can be performed manually with the help of that table and it can also be performed automatically by SPSS. Here is how we use a random number table to choose a sample.

To select a simple random sample, start with the following steps.

1. **Make a list of all the units of the population, and number them as follows:** 001, 002, 003, etc. You want *all units to be numbered by the same number of digits*, which is why we are throwing in additional zeros. If we have less than a hundred individuals, we only need two digits, and the numbering will be: 01, 02, 03, … etc., up to 99. We can also start at 00, 01, 02, 03, … etc., up to 99. If we

start with 01 and go up to 99, we can include
99 individuals in our sample. But if we start
at 00 and continue up to 99, we can include
100 units, not just 99 of them. If we have
more than a hundred individuals but less than
a thousand, we can number them as: 001,
002, 003, …, etc., up to 999, or again we can
start at 000 to be able to include 1000 units,

> A **digit** is a numerical
> character used to form a
> number. The number 65 is
> formed of two digits, 6
> and 5. The number 782 is
> formed of three digits: 7,
> 8, and 2.

not just 999. The idea is to have a numbering system where all units are labeled
by numbers that have *the exact same number of digits*. We will choose the sample
from that list, using the unit numbers. This lists of units from which we choose a
sample is called the **sampling frame**. Ideally, the sampling frame consists of the
whole population, but this is not always possible. Sometimes the sampling frame
will not be all inclusive, but we always try to make it as complete as possible.

Example

Suppose, for example, that you want to select a random sample of 100 households from a small town. You do not have the list of all households in the town,
but you do have the telephone book of that town. You will use the list of telephone numbers in the phone book to select the sample. The population in this
case is the set of all households in the town, and the list of telephone numbers
in the phone book is the sampling frame. It is not a perfect sampling frame,
since it does not coincide exactly with the list of households: some households
are not listed in the phone book, while others may be listed more than once if
their members want to be listed separately. But it is a reasonably good sampling
frame for market research. It would not be a good sampling frame for rigorous
social surveys.

2. **Determine how many units you need in your sample.** In order to provide the
   basis for a sound generalization, samples should not be too small. A bare minimum is 30 individuals for a simple random sample, but a hundred individuals is
   certainly better. If the population consists of various groups that may differ on
   the variables that we are measuring, we will need a larger sample. In the
   chapter on estimation, we will see that the larger the sample, the smaller the
   margins of error. We will then be able to discuss again the question of the size
   of a desired sample.

3. **Using a table of random digits, choose a random sample as explained below.**
   Suppose that you have less than a hundred units, so as to need only two digits to
   number them. Take the table of random numbers given in Appendix 2. The table
   can be entered at any row or at any column, and in any direction. Choose any

row or any column to start reading the table, and determine whether you want to read it from right to left or from left to right (for rows), or from top down or from bottom up (for columns).

Let us say you choose the **fifth row from right to left**.

You move along the row from right to left, reading two digits at a time. Whenever you meet a number that corresponds to a unit in your list, you note it down; it is part of your sample. If you only have 72 units in your sampling frame, numbers such as 86 would simply be ignored, as no unit corresponds to them. You stop when you have reached the required number of units in the sample, ignoring the second occurrence of any number.

Example

Suppose your sampling frame contains 54 units, and you want to select 6 of them. You have chosen the following row from the table of random digits and the direction you have chosen is from right to left:

39198 23832 72291 35735 73716 76410 67487 80109 35775 18835

Going *from right to left*, you encounter the following two-digit numbers:

53 88 15 77 53 90 10 87 84 76 01 46 76 17 37 etc.

You want to keep only the numbers that correspond to a unit in your sampling frame, and simply drop the other numbers. We have rewritten the list of two-digit numbers, with the numbers that correspond to your sample underlined, the others being crossed out:

<u>53</u> 88 <u>15</u> 77 90 <u>10</u> 87 84 76 <u>01</u> <u>46</u> 76 <u>17</u>

Notice that we have kept all the numbers under 54, because they correspond to units in the sampling frame. The number 53 has been kept only once, and the second occurrence has been dropped. We stopped when we reached 6 units. So the sample chosen includes the 6 units numbered 53, 15, 10, 01, 46, and 17.

Your simple random sample has thus been selected.

Selecting a random sample with SPSS   SPSS can do the random selection for us, as indicated in the lab session on random sampling. Follow the instructions in Lab 7 to learn how to select a random sample with SPSS.

SYSTEMATIC SAMPLING

There is another type of probability sample which can be easily selected. It is called a **systematic sample**. To select a systematic sample we list the units as we did above for the simple random sample (SRS). We select the very first unit at random, and then we go down the list of units (that is, the sampling frame) at *regular intervals* in order to get the required number of units in our sample.

Example

Let us say we have 300 units in our sampling frame, and we want to select 15 of them. We divide 300 by 15, and get 20 (if this number is not a whole number, round it *down*). That means we want to have exactly one out of every 20 units. We select a single unit at random from the first 20, then we take every 20th unit in the remaining part of the list. With this procedure we end up with 15 units selected, but if the number of units in the sampling frame is not evenly divisible by the size of the desired sample, a number of units towards the end of the list will have no chance of being selected. This could be compensated by selecting at random one of the units left out and adding it to the sample. Thus, when the size of the sampling frame is not a multiple of the size of the desired sample, systematic samples do not fully comply with the condition that all units have the same probability of being selected.

If 15 units out of 300 are selected by various individuals using the table of random digits, it is expected that these individuals will obtain samples that are not identical because they are likely to use different lines or columns in the table of random numbers to select the first unit in the systematic sample.

CLUSTER SAMPLES

It is sometimes very costly to conduct a study using a simple random sample, as the units chosen may be spread out over a vast territory. This is why the technique of cluster sampling is used. To obtain a cluster sample, you first group your population into *clusters*, that is, small subgroups. For example, in a survey in an urban area, the clusters could consist of all people living in a street block (the smallest set of houses surrounded by streets on all sides). Or a cluster could be all people living in a stretch of street without intersections. In a survey conducted in rural areas, a cluster could be a village.

Once the clusters have been determined, you could select a *simple random sample of clusters*. This means that what you select is a whole cluster, not an individual. Then, you interview every person in the cluster. Or you select at random, in every cluster, a fixed number of people to be interviewed. In this way, your sample is still a probabilistic sample, but it is not a simple random sample.

Clustered samples are generally used for surveys on consumption in large cities. They are also used by national statistical agencies in their regular surveys.

STRATIFIED RANDOM SAMPLES

Sometimes the population consists of various subgroups, and we want to be sure that each subgroup is adequately represented in the sample. For instance, we may have men and women in a given proportion, and we want to make sure the sample reflects that proportion very precisely. Or we may have different geographical areas, and we want them all represented in the same way in our sample. Sometimes the subgroups are defined by a combination of two variables. If you have men and women, and three levels of income, you may want to have all six groups adequately represented:

- Men with a high income
- Men with a medium income
- Men with a low income
- Women with a high income
- Women with a medium income
- Women with a low income.

In all the cases mentioned above, we may want to select a **stratified random sample**. The word *stratified* means that we divide the population into subgroups (called *strata*; the singular is *stratum*) according to some variables (such as sex or geographical area of residence or income or some combination of them), and then we select a random sample *within each* stratum.

We now have two choices: we may want the number of units within each stratum to be *proportional* to the size of that stratum. If the number of *Men with a high income* is twice as big as the number of *Women with a high income* in the population, we may want the sample to reflect that same proportion. A sample reflecting faithfully such proportions is called a **proportional stratified random sample**. Statistics calculated on a proportional stratified random sample can be readily generalized to the population, and the parameter is estimated to be equal to the observed statistic (plus or minus a margin of error, as we will see in a subsequent chapter).

In some cases, it is not desirable to have a proportional stratified random sample. Here is an example of such a situation.

Suppose that we conduct a study on working conditions in manufactures in a certain city. We want to stratify the sample on the basis of the size of the manufactures. Very large manufactures (more than 500 employees) are less numerous than medium-sized ones (50 to 500 employees). We may want to take 10% of the medium-sized manufactures, but if there are only 3 large ones in the city, we cannot take 10% of that number … We will then study one or may be two of the large manufactures. Our sample will therefore not be proportional. We call it a **non-proportional stratified random sample**. Statistics calculated on such samples will have to be corrected by the method of weighted means explained in Chapter 3 to compensate for the unequal proportions of the various strata. For instance, to calculate some average, we will compute the *weighted means* rather the means, the weights being calculated to reflect the relative importance of the various groups of manufactures.

Non-Probabilistic Samples

While probabilistic samples are the best guarantee that the sample will tend to yield accurate results that can be generalized to the whole population, it is sometimes impossible (or too costly in terms of time or money) to get a probabilistic sample. In these cases, non-probabilistic samples are used, but you should be aware of the fact that results based on such samples are not as reliable. They tend to be **biased** because the units selected may share some characteristics that set them apart from the whole population, from which they may differ significantly. When the sample is not a probabilistic one, we cannot assess the risk of error as accurately as we do with probabilistic samples. We can distinguish the following types of non-probabilistic samples.

QUOTA SAMPLES

Quota samples share some characteristics with stratified random samples, but they differ from the latter in that they are not probabilistic. Like stratified random samples, they include various groups in the same proportions as in the general population. Such groups are defined on the basis of certain criteria, for instance socio-economic variables such as sex, age, level of income, place of residence, mother tongue, ethnic origin, etc. The sampling design specifies how many of each group you need. For instance, if you are doing a survey in Canada, and your general population includes 30% of Canadian residents born outside of Canada, you want to obtain a sample that includes 30% of individuals born abroad, thus reflecting the proportion found in the population. For each one of the basic criteria, you determine how many people you need for each of the categories of the variable.

A researcher doing a survey on a quota sample is free to select in any convenient way the first few interviewees. But as you progress in the constitution of the sample, you will have to be careful to select persons having very specific characteristics, in accordance with the sampling design. For instance, if you have already interviewed enough people who are male, born in Canada, and having stopped their education at high school, your sample design may request that you find a female, born abroad, and having a university degree. Here is an example of how a quota sample is selected.

Example: selecting a quota sample

You want to select a sample representing 10% of the students in a college, in such a way that it reflects the population on the basis of three criteria: Sex, Age, and Program of Studies. Let us say we have a student population of 5332 individuals, distributed as shown in Table 6.1.

If you want the sample to include10% of the student population, you must have 533 individuals. Since the female students constitute 55% of the population, they must also constitute 55% of your sample, that is $(0.55 \times 533) = 293$ students. The male students will constitute 0.45 of 533, which is 240 students.

Table 6.1

Variable	Categories of the variable	Number of students	Percentage of students in each category	Number of individuals in the sample
Sex	Men	2402	45	
	Women	2930	55	
Total		**5332**	**100%**	
Age	16 or 17 years	889	16.7	
	18 years	1596	29.9	
	19 years	1238	23.2	
	20 years	685	12.8	
	21 years or more	924	17.3	
Total		**5332**	**100.0%**	
Program	Natural and Health Sciences	1121	21.0	
	Soc. Science and Commerce	1753	32.9	
	Fine Arts	180	3.4	
	Computer Science	567	10.6	
	Office Systems Technology	1017	19.1	
	Business Administration	694	13.0	
Total		**5332**	**100.0%**	

You must do a similar computation for each variable. To find out how many people you need in a given category, multiply 533 by the proportion represented by that category. Do this computation on your own and write the results in the last column of Table 6.1. You should end up with the following numbers.

In recruiting the 533 individuals in your sample, you must make sure that you have:

1. 240 men and 293 women (total = 533);
2. 89 of them must be 16 or 17 years old (that is 16.7% of 533), 160 of them must be 18, 123 must be 19, 69 must be 20, and 92 must be 21 or over (total = 533);
3. 112 of them must be in Natural and Health sciences, 175 in Social Science and Commerce, 18 in Fine Arts, 57 in Computer Science, 102 in Office Systems Technology, and 69 in Business Administration, for a total again of 533 individuals.

You must realize that these are not three different groups of students: it is one single sample of 533 students, but the individuals in that sample must belong to the various categories of the three variables in the proportions and numbers indicated above. This is a quota sample.

While quota samples are not probabilistic, they do reflect the characteristics of the population, and they are widely used in market research. Generalizations based on them are not as reliable as those based on stratified random samples, because margins of error cannot be determined accurately in quota samples. But they can be used whenever high accuracy is not an issue.

CONVENIENCE SAMPLES

Convenience samples are chosen haphazardly among a set of units which are readily available, but they are not random samples, and they usually contain important biases. Suppose that a journalist stands at an intersection downtown, and asks people, haphazardly, what they think about some issue. He or she may have the impression that the subjects have been chosen at random, but this is not the case: the people who are likely to pass by that intersection are not a representative sample of the population. They may include a larger proportion of people with certain characteristics than is found in the population, and the journalist may not be aware of that. Therefore, results obtained on the basis of convenience samples should not be considered accurate. Opinions gathered in this way are therefore not representative. They may be very informative about the **range** of opinions found in the population, but **not** about the **proportions** in which such opinions are found in the population.

JUDGMENT SAMPLES

Judgment samples are not based on randomness. They result from judgment that makes us believe that analysis of certain specific units is more likely to give us a better idea of the problem we are examining. Suppose for instance you want to study the problem of violence in high schools. You may judge that a couple of high schools in the city are more interesting to study because of the level and kind of violence that is prevailing in them. If you include some schools specifically in your sample because you know that they will be interesting to study, you are constituting a judgment sample. The results that you will get from studying them are not generalizable, because you cannot estimate the margins of error of your measurements. However, such a study could be instrumental in identifying *patterns* and *processes*, in understanding the problems from within. Such a study would be more qualitative than quantitative. After such a study is completed, you may want to follow up the issue with another study based on a random sample of students or of schools, to see the extent to which the various factors of violence that you have identified are prevailing in the general student population.

Thus, we can conclude that studies based on a judgment sample are more often qualitative than quantitative. But even if they include quantitative aspects, their purpose is to identify relevant factors, not to generalize the frequency of their occurrence.

SAMPLES OF VOLUNTEERS

The last type of non-probability samples are samples of volunteers. Samples of volunteers are composed of people who respond to a general appeal, without any selection. For instance, a newspaper may conduct a poll with its readers, asking them what they think about a given issue. The questionnaire is published in the newspaper, and readers are invited to respond. The people who respond constitute a sample of volunteers. Such samples are highly biased and should not be taken too seriously, for three reasons.

The first reason is that the readers of a given paper are not a representative sample of the population. Each paper has a certain style that attracts some categories of the population (for instance, people with more or less schooling, more or less conservative, etc.). The readership itself implies that the answers will be biased in a direction that may be difficult to assess.

The second reason is that people who respond may share some common characteristics that make them more likely to respond, thus increasing the bias resulting from the readership of the paper. There may be a higher proportion of people who are retired, for instance, and who can afford to take the time needed to answer the questionnaire. Thus, the answers will reflect the opinions of that particular group of people more than they reflect the opinions of the population in general.

The third reason is that when the issue is politically or morally sensitive, interest groups may mobilize and flood the paper with letters, or answer massively a questionnaire in a way that biases the results even further. This is particularly true of radio programs where people phone in to voice their opinions.

A classical example of this kind of situation is provided by the Hite reports published some years ago. The two reports on female sexuality published by Shere Hite in 1976, then in 1987, pretended to give a portrait of the sexual life of American women. The questionnaires were widely circulated in popular magazines, as an annex to the first book, and through the mailing lists of a number of associations. For the second book, a total of 7239 filled questionnaires had been returned to the researcher. This may be a large number but it does not make it a representative sample, especially when one considers that a total of 119,000 questionnaires had been in circulation. Not only was the response rate very small (close to 6%) but the self-selection of women who decided to answer the questionnaire introduced an important bias in the report. As a result, the report cannot be considered a faithful picture of reality. However, the report did have a unique and original value in that it showed that a certain *range* of sexual practices were more widespread than expected. But how widespread? The report could not answer that question because the sample of people who answered it is a sample of volunteers.

Errors of Measurement

In spite of all our efforts to be as precise and as rigorous as possible, the measurement of any variable includes errors. The total error in any measurement can be broken down as shown in Figure 6.1.

Here is an explanation of each type of error.

Errors of Observation

This type includes all errors that are not due to the choice of the sample. Any error resulting from a poorly formulated question, or from a high rate of non-returns, or from a wrong operationalization of the concepts, or from a mistake in transcribing the results is an error of observation.

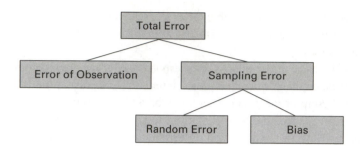

Figure 6.1

Sampling Errors

These are the errors that result from the very operation of sampling. By its very nature, a sample may differ from the population simply because it does not include every individual. We minimize this error by taking a random sample. But even a random sample may differ from the population, thus causing an error in our measurement. We call this type of error a **random error**. Thus, a random error results from the choice of a random sample, even when the sample has been chosen in accordance with all the rules of random selection. *Such errors can be calculated*, and this is what we will learn in the chapter on statistical inference.

But a sample may also include a **bias**, that is, a systematic error (of which the researcher may not be aware) that distorts the results. Such errors can be minimized by a careful choice of the sample. If the sample is not a probability sample (see the section above), chances are that the bias is going to be important. The only type of non-probability sample that reduces the bias is the quota sample. Errors that are due to bias cannot be calculated, and they distort the results and diminish their value.

Summary

We are now able to formulate a precise definition of the **sampling design**: it is a procedure for selecting a sample that specifies:

- the type of sample to be used;
- the number of units to be selected in the sample as a whole and in the various strata if needed;
- the method for choosing the units.

There are basically two types of sampling design: those based on probability and those not based on probability. The basic type of probabilistic sample is the simple random sample, which can be chosen with the help of a table of random numbers. Most statistical packages include procedures for automatically choosing a sample at random from the list of units included in a file.

Probabilistic samples and quota samples tend to reduce the error resulting from bias of the sample. With probabilistic samples, we can compute the expected margin of error (this will be seen in a subsequent chapter), but we cannot do it for a quota sample and for all other non-probabilistic samples. In all kinds of samples there are possible errors of observation that do not depend on the operation of sampling itself; such errors can be reduced by following carefully the rules of research methodology.

Keywords

Sampling	Sampling design	Sampling frame
Probabilistic samples	Simple random samples	Systematic samples
Non-proportional stratified random samples	Stratified random samples	Proportional stratified samples
Cluster samples	Non-probabilistic samples	Quota samples
Convenience samples	Judgment samples	Volunteers samples
Error of measurement	Total error (of measurement)	Sampling error
Random error	Bias	Error of observation

Suggestions for Further Reading

Bechhofer, Frank (2000) *Principles of Research Design in the Social Sciences*. NY: Routledge.
De Vaus, David A. (1991) *Surveys in Social Research* (3rd edn). London: UCL Press.
Wonnacott, Thomas H. and Wonnacott, Ronald J. (1997) *Introductory Statistics* (3rd edn). New York: John Wiley and Sons.

EXERCISES

6.1   Suppose you want to choose 7 units in a sampling frame of 257 units. Indicate below the details of the procedure.

Numbering system for units:

_____

Choose the row or column in the table of random numbers, and the direction of moving along it:

Row number chosen: _____ or Column number chosen: _____ .

Direction of movement: _____

Indicate the first 10 numbers you encounter (whether they are selected or not; each number is made up of 3 digits):

_____ _____ _____ _____ _____ _____ _____ _____ _____ _____

Indicate the numbers of the 7 units selected (you may have to go further down in the list than the 10 numbers listed above):

___ ___ ___ ___ ___ ___ ___

What you have obtained is a simple random sample of size 7.

6.2 You want to conduct a study on the working conditions of your peers in your college. You can't have the list of all students enrolled in the college, but you are able to get the list of all students enrolled in the Methodology courses. So, you select at random 45 of them.

What is the population?

_____

What is the sampling frame?

_____

What kind of sample is that? (Refer to the typology of sampling designs shown in the manual.)

_____

6.3 For your research project you need to select a stratified random sample consisting of a total of 70 students. The sampling frame is the list of students enrolled in seven classes, with a total of 210 students. You want the sample to be stratified on the basis of two variables: Sex and Program of study (Commerce or Social Science). For the whole group, the frequencies of these variables are as follows:

	Social Science	**Commerce**	**Total**
Men	39	60	99
Women	69	42	111
Total	108	102	210

Describe all the steps you would go through, indicating how many students would be in each stratum of your sample and what method you would use to select them.

6.4 For each example below, indicate what kind of sampling design is involved.

(a) You select haphazardly 10 of the people you know to answer a question-naire about their preferences in beauty products.
(b) You select randomly 5 tables in the cafeteria, and then you interview every person sitting at each of these 5 tables.
(c) You publish your questionnaire in the student newspaper and you ask readers to fill it up and send it to you.

(d) You ask friends to help you pass a questionnaire. Each interviewer must pass 10 questionnaires. You tell them that they can interview anybody they like provided their sample contains:

6 women and 4 men;
4 Francophones, 4 Anglophones, and 2 Allophones;
4 in Commerce, 3 in Social Science, and 3 in the IB program.

7

DATABASES ON SOCIAL STATISTICS

The purpose of this chapter is to gain some familiarity with the databases on social statistics that are available online through the Internet, so as to be able to find and retrieve social statistics that could be found on these sites. The general principles of organization of such data will be presented here. This presentation is to be complemented by Lab 8 which includes some practical tips.

After studying this chapter, the student should know:

- how to use printed and electronic databases;
- the kinds of data that is found on such sites (aggregate or individual) and their formats;
- how to explore the websites of various national and international agencies;
- how to find specific quantitative information on these websites.

Statistical databases are archives of primary data, compiled and maintained by specialized organizations. They usually involve large quantities of data files that span a wide spectrum of questions, over a long period of time. Examples of such specialized organizations are provided by the official national statistics institutions of most countries. These institutions tend to be governmental or para-governmental, and they have a very specific mandate: that of compiling statistical information about various aspects of the life of the nation, and of making it available to decision makers, to researchers, and to the general

Recall that **primary data** is data that is published by the organization that collected it in the first place. For example, the data published by most official national statistical institutions (such as the UK Office of National Statistics, the US Census Bureau or Statistics Canada) is primary data. When used by a researcher and published as part of a study, this data becomes secondary data because the researcher is not the one who collected it. Secondary data tends to be partial, because it is selected to answer a particular research question. Therefore it is less comprehensive than the primary data it was part of. Primary data is usually considered to be more reliable than secondary data. When given the possibility to do it, it is always better to check thte accuracy of the data at its primary source.

public as well. We are going to learn how to access the primary data compiled by such institutions, as well as reports and analyses done by these organizations.

There are four types of document you can find in an online statistical database:

- tables containing aggregate data;
- short descriptive reports highlighting the important statistical features of a situation;
- in-depth analytical report, containing data as support for the conclusions;
- data sets of individual data (usually primary data).

A given website may not include all four types, and if it does, access may not be permitted or free to everybody.

Aggregate data is presented in tables, and it is the easiest to find. But the units used to aggregate the data may not always suit the researcher. For instance, the data presented in most United Nations' and other international organizations' sites uses the individual country as a unit of analysis. This may be useful for some research purposes but not others. National sites will present a lot more social indicators than the websites of UN bodies, and they may break down the unit of aggregation of the data by city or by region, but such sites also make available data files where the unit of analysis is the individual.

Individual data is more complex. In order to be interpreted correctly a data file must be accompanied by several meta-data files: files that explain how to interpret the data, such as the codebook, or the questionnaire itself that was used to produce the data. A brief explanation of how these data files are presented and what meta-data files contain is presented in Insert 7.1.

INSERT 7.1   Data Files Can Come in Three Types of Formats

ASCII data is simply lines of text. To decode it, one must know exactly how the data is stored: whether the data relating to one statistical unit is spread over one line or more; the exact location of the data that refers to a specific variable; the questionnaires that were used to compile the data; etc. This information must be clearly stated in the meta-files, otherwise the data cannot be interpreted.

CSV files are files where the values referring to different variables are separated by a comma. They come with the suffix .csv which stands for 'comma separated variables'. This data must also be accompanied by meta-files that explain how to interpret it. The character that delimits the various variables is generally a comma, but it could also be a tab character, or a semi-colon, or some other specific character. The procedures for importing such data into a statistical package such as SPSS will differ from the procedure used for ASCII data, but both types need the meta-files to be interpreted correctly.

Sometimes the data comes in **simple tables** that can be readily transformed into a spreadsheet, but there too it needs to be accompanied by explanations.

On some of these sites, the visitor to the website is not allowed to download a huge data file, but he or she is asked instead to select a limited number of variables among those that were used in a data file. Only the data that relates to these variables will be downloaded.

The exact sequence of moves to be performed to complete this operation depends on the site and on the software used. Every site includes instructions on how to access its databases and how to download them.

As for the meta-files, they usually include the following:

- the codebook, indicating the variable and value labels, the data format, and the data position;
- a data dictionary, that lists the variables and their labels in alphabetical order;
- the questionnaire that was used to produce the data;
- technical explanations on the survey itself, and on its methodology, including the sampling design used to collect the data;
- any other information that is needed for a correct interpretation of the data.

Sometimes, the meta-files will include a syntax document that will, when downloaded and run, automatically transform the raw data files (ASCII or CSV) into the format of a well-known statistical software such as SPSS or SAS. In this case, the reader downloads the data file and the syntax file (sometimes called 'SPSS card' or 'SAS card'), then runs the syntax file. This operation creates a data file in the format of the corresponding statistical software.

Most of the information available on national websites centers about three domains: the economy, demography, and social indicators (health, education, justice, the family). Some sites also include statistics about the political system and political behavior, and some include environmental information as well.

The statistical data made public by such national institutions can generally be found in three formats:

- on-line data (on the website of the institution);
- printed reports;
- CD-ROMs and disks.

You will find that it is possible to order (usually for a fee) CD-ROM products and printed reports from these institutions, and even custom-made data files on diskettes, which are assembled for a specific objective and contain far more detailed and specific data.

In addition to the official national statistical institutions mentioned above, there are international bodies who also make available statistical data. Sometimes these international bodies are the source of the data themselves, and sometimes they compile it from data provided by individual countries. In addition to the sites of the UN family of institutions, such as UNESCO, UNICEF and UNDP (the United Nations

Development Program), the UN has a general statistical division hosted at http://www.un.org/Depts/unsd/. The site map of this location should be consulted as it contains all the links that are available from this site, and it would allow the reader to access the required data in the most efficient way. These UN sites have a tremendous wealth of data, but the statistical unit used for this data is evidently the country, and one cannot find in these sites data that concerns smaller units. However, one can find references to studies, usually sold for a fee, that contain analyses and secondary data on sub-national units. Sometimes the full text of such studies is available online and can be either read directly or downloaded.

Some academic institutions such as the Inter-university Consortium for Political and Social Research (ICPSR; http://www.icpsr.umich.edu/), based at the University of Michigan in the US, have made available a large collection of data files. The International Association for Statistical Computing (http://www.stat.unipg.it/iasc/Misc-stat-official.html) contains direct links to many international associations and to a seemingly exhaustive list of existing national statistical sites (over one hundred sites).

When looking for particular information, there are two ways of finding it: either by the table of links of the website, which functions like a table of contents and which leads us from general topics to increasingly detailed ones, or through the search function of the website, which can turn out documents that could not be found through the links of the site. Usually, all tables of numerical data will be available through the links of the web page, and studies and reports which include secondary data will be found only through the search function.

In addition to this site-specific search, the option to do a web-wide search is always available, and the search engines increasingly powerful. A search with Google (www.google.com), for instance, can produce an incredible amount of relevant and well-focused material. There are at least two problems with these searches. The first is that they tend to produce far more links than needed, of which only a small proportion is directly relevant. The second is that they do not distinguish between reliable and non-reliable sources: the reader must make his or her choices according to some criteria. The first problem is not too complicated to solve, as we can narrow down the search with more keywords. The second problem is more tricky. Official institutions are generally reliable, but they also have their 'blinds spots' and ignore sometimes important issues that are taken on by smaller, alternative sources of information which are sometimes rigorous and sometimes not.

Example

We did a search on Google with the key word 'juvenile deliquency'. Google spotted the typing error and had the kindness of stating: 'Do you mean *juvenile delinquency*' and went on to list 1760 links.... When we added the word 'statistics' to the search terms, the number of links was reduced to 436 links.

We will now examine each type of information mentioned above, and we will illustrate it with data taken from one or other of the websites mentioned.

INSERT 7.2   A Technical Note on Searching With Keywords

A good search on the web (as in any other type of library search) must rely on a clear list of search terms, which may be complemented as we go. Some databases provide their own Thesaurus. If you want to consult the database on a term like 'youth violence', for instance, you first consult the Thesaurus, which would suggest alternative or additional terms. Some Thesauruses would suggest, for a search on youth violence, the alternative term 'juvenile delinquency'. Failure to check the Thesaurus means that you may miss important documents that are available, but that were missed because of an inadequate search keyword.

If no Thesaurus is available in a given database, you must establish yourself a list of search keywords that are likely to be used by the database. A thorough search would use all terms of the list. It is also a good idea to keep track of which terms were used in which databases, to avoid duplication and to make sure that all possibilities were considered. You should keep these remarks in mind when looking for statistical information on the web or in published catalogs.

Illustration of a Data Search on Selected Sites

We will illustrate the way we can go about exploring these sites by doing it on a few examples.

The UN Statistics Division

The home page of the UN Statistics Division (http://www.un.org/Depts/unsd/) explains what the division is, and offers a number of links such as Economic Statistics, Demographic and Social Statistics, etc. If we click on the link Demographic and social statistics, we get the web page shown in Figure 7.1. Take a look at the list of links. It gives you an idea of the kind of data you may find. If we now click on Social Indicators, we get a page that includes the following list of links:

Social Indicators

- Population
- Youth and elderly populations
- Human settlements
- Water supply and sanitation
- Housing
- Health
- Child-bearing
- Education
- Literacy
- Income and economic activity
- Unemployment

Statistics Division   Search the site [   ] GO

| HOME | STATISTICAL DATABASES | PUBLICATIONS | METHODS AND CLASSIFICATIONS | STATISTICAL COMMISSION |

Millennium Profiles   **Demographic & Social**   Energy   Environment   Industry   National Accounts   Trade

▼ Demographic, Social
and Housing Statistics

Population and vital
statistics report for
subscriber access
Social indicators
World's Woman 2000
Disability statistics
International economic and
social classifications
International Migration
Statistics
Population and housing
census dates, 1985-2004
World Population and
Housing Census
Programme
Vital Statistics
City groups on statistical
methodologies
Development indicators
Time-use surveys

Demographic, Social and Housing Statistics

Demographic statistics on-line

The Population and Vital Statistics Report (available to MBS subscribers only), a quarterly journal in English, gives the latest estimates of world and continental population, and for 229 countries or areas of the world. The information is derived from each country's latest national census and birth and mortality figures.

Social statistics on-line

The Demographic Section maintains a database of social indicators which monitor the results of recent major United Nations conferences on children, population and development, social development and women.

The World's Women 2000: Trends and Statistics is a statistical source-book analysing how women compare to men worldwide in the areas of families, health, education, work, population, and human rights and politics.

The Disability Database is a response to the World Programme of Action Concerning Disabled Persons by implementing a convenient statistical reference and guide to the available disability data.

Demographic statistics methods

The Division also provides links to reference sources for International Economic and Social Classifications.

Migration statistics tracking the flow of people can befound on line. The Statistics Division also publishes Recommendations on Statistics of International Migration Revision 1 on identifying and measuring migration flows.

Figure 7.1   **Source: United Nations Statistics Division. © United Nations**

If we want to see what kind of data we get, we click, say, on Water supply and sanitation. Table 7.1 shows the first few lines of the table we get.

You notice that the units of analysis here are countries. All countries are listed in alphabetical order. When the data is not available, dots are written instead. The titles of the columns are hyperlinks, and when you click on one of them you can read a technical note about the way that variable was measured.

Hundreds of tables are thus available. The *Report of the Expert Group on the Statistical Implications of Recent Major United Nations Conferences*, listed in the Social Indicators home page, discusses among other things all the indicators that should be included, by sector, and it lists them in the appropriate section. This site can therefore provide us with a lot of social indicators if you need these statistics by country.

Table 7.1 **Indicators on water supply and sanitation**

	% of population with access to improved drinking water sources			% of population with access to improved sanitation facilities		
Source: World Health Organization and UNICEF.						
	2000			**2000**		
	Total	Urban	Rural	Total	Urban	Rural
Afghanistan	13	19	11	12	25	8
Algeria	94	88	94	73	90	47
American Samoa	100	100	100	...	...	...
Andorra	100	100	100	100	100	100
Angola	38	34	40	44	70	30
Argentina	79	85	30	85	89	48
Australia	100	100	100	100	100	100
Austria	100	100	100	100	100	100
etc ...						

But the Search function of this site is rather crude: it finds all documents containing one or other of the words used in the search phrase. For instance, if you type 'Juvenile crimes' you will get all documents containing in their text either the word 'juvenile' (none of the documents includes this word) or the word 'crime'.

As was mentioned before, searches can also be performed with a search engine on the whole World Wide Web.

Internet for Social Statistics

Internet for Social Statistics (http://www.sosig.ac.uk/vts/social-statistics/) has been written by Robin Rice of the Edinburgh University Data Library. It is a very well crafted site that guides readers through a list of sites and offers a tutorial on social statistics on the Internet. It also proposes a critical approach to the subject and raises quite adequately the question of rigor and reliability of the data.

Inter-university Consortium for Political and Social Research

This is a site that deserves to be thoroughly explored (www.icpsr.umich.edu/). A part of the links in the home page is shown in Figure 7.2. We will focus on only two links in its home page, the GSS Directory and the Quick Data Links.

The GSS Directory site includes all the available information about the General Social Surveys conducted in the US since 1972: the data files, and a large collection of meta-data files.

That site also contains indexes for the studies and reports made about the GSS surveys, some of which are found on the site with their full text. When, for example, we conducted a search with the keywords 'sexual permissiveness', we obtained 22 results, of which 9 were reports, and most others were bibliographic entries. The

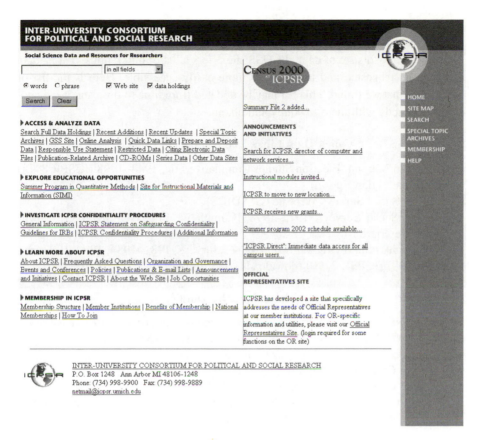

Figure 7.2 **Source: Inter-university Consortium for Political and Social Research. © ICPSR**

first report, for example, turned out to be: *Attitudes Towards Sexual Permissiveness: Trends, Correlates, and Behavioral Connection*, by Tom W. Smith, National Opinion Research Center, University of Chicago, June, 1992. The full text of the report was available online.

As for the data resulting from the General Social Surveys, not only can you access it, but you can also access the questionnaires themselves, the lists of topics that were tackled, and the list of variables that were included and their definitions. The variables used in one or other of these surveys (which are organized in cycles, every cycle being a set of surveys that covers a general topic) are listed in various forms: alphabetically, chronologically, by theme, etc. When you visit these various indexes, you can pick the variables of your choice. After you have finished picking the variables you want to examine, you extract them. The site engine then builds a compressed file that you can download on your computer, and which includes all the individual data that relates to the variables that were picked. You must also download some of the meta-data files that will allow you to read the data and manipulate it. If you download, for instance, the SPSS Card file, you simply have to run the syntax included in this file and SPSS will build the data file for you, which you can then save and analyze as you wish. Recall that all these files are downloaded in compressed format and you have to extract (or unzip) them.

In addition to this set of data, the GSS site offers a list of quick links, that is, data that is often requested and used, and is more easily accessible. Here is the list of quick links that we found. This list may be updated from time to time and it may not coincide exactly with the one you would obtain.

Quick Data Links listed on the GSS site

- American National Election Studies Cumulative File, 1948–1998 (ICSPR 8475) (downloads | description)
- General Social Surveys WebSite (GSSDIRS)
- General Social Surveys, 1972–1998: [Cumulative File] (ICPSR 2685) (downloads | description)
- World Values Surveys & European Values Surveys, 1981–1984, 1990–1993, and 1995–1997 (ICPSR 2790) (downloads | description)

Criminal Justice Data

- United Nations World Surveys on Crime Trends and Criminal Justice Systems, 1970–1994: Restructured Five-Wave Data (ICPSR 2513)
- Homicides in Chicago, 1965–1995 (ICPSR 6399)
- Uniform Crime Reporting Program Data: [United States]
- National Crime Victimization Survey (ICPSR 6406)
- For additional information on these or similar studies, please visit the National Archive of Criminal Justice Data (NACJD).

Substance Abuse/Mental Health Data

- National Household Survey on Drug Abuse, 1998 (ICPSR 2934)
- Monitoring the Future: A Continuing Study of American Youth (8th- and 10th-Grade Surveys), 1998 (ICPSR 2752)
- National Comorbidity Survey, 1990–1992 (ICPSR 6693)
- For additional information on these or similar studies, please visit the Substance Abuse and Mental Health Data Archive (SAMHDA).

Education Data

- Common Core of Data: Public School Universe Data, 1996–1997 (ICPSR 2823)
- Schools and Staffing Survey, 1993–1994 [United States]: Revised Version (ICPSR 2748)
- For additional information on these or similar studies, please visit the International Archive of Education Data (IAED).

Aging Data

- Panel Study of Income Dynamics, 1968–1995 (ICPSR 7439)
- Census of Population and Housing, 1990 [United States]: Public Use Microdata Sample: 5-Percent Sample (ICPSR 9952)
- For additional information on these or similar studies, please visit the National Archive of Computerized Data on Aging (NACDA).

Health and Medical Care Data

- Practice Patterns of Young Physicians, 1997 (ICPSR 2829)
- Community Tracking Study Physician Survey, 1996–1997 (ICPSR 2597)
- For additional information on these or similar studies, please visit the Health and Medical Care Archive (HMCA).

We will now examine the way official national statistics institutions structure their published data. We will illustrate our discussion with the case of Statistics Canada, as it is reputed for its high standards. We will examine in some detail the documentation available at Statistics Canada. The details of the organization of these files vary across the different official national statistical institutions, but the basic principles of organization of these files remain the same, and the familiarity gained with the Statistics Canada documentation will easily be transferable when you come to examine other national sites.

Printed Reports of Statistics Canada (or StatCan)

StatCan produces several publications where its data is published and analyzed. A large number of these publications are available at most academic libraries in Canada. Because these publications are numerous and varied, their list is published every few years in a catalog. The 1994 catalog is more detailed than more recent ones. The StatCan publications are given a call number and are organized on the shelves by these call numbers. In order to find data on a particular topic, you must first refer to the StatCan catalog. The first part of the catalog lists all publications by catalog number, in an ascending order. The publication's name is given and the publication content is briefly described.

A very useful aspect of the catalog is its **index**, which is the second part of the catalog. If you want to do a research on family violence in Canada, for instance, you look under the word Family in the index, then you locate a sub-entry that describes the topic more precisely. It will list the call number of the publication, and you can find it in the appropriate drawer or shelf. More generally, you must start every search for documents on a given topic by establishing a **list of keywords** that are related to that topic, as was mentioned earlier in this chapter (see Insert 7.2). You then search in the index for these keywords, and you select among the publications that you have found the ones that address your topic precisely.

StatCan publishes two broad categories of data: the raw data itself that it has collected, and short or long articles that analyze a specific question, giving only the data needed to support the conclusions or to illustrate what has been observed.

Raw Data

Because the Census data is far too voluminous to be published in a single report, the raw data collected by StatCan is published in the form of frequency tables in various thematic reports, usually kept in the Reference section of academic libraries. For instance, one volume will contain the numbers of people living in all municipalities, by Sex, or by Age groups. Increasingly, this data is also available online or on CD-ROMs or on disks that can be purchased. It is expected that at some point, this data will be available in libraries only in electronic formats.

Analyses

The analyses are published in various publications, which differ in their degree of specialization. You can find such studies on all aspects of the social and economic life in Canada, on the social behavior of individuals, on demographic issues, on health, on accidents, on the industry, on trade, on mines, on fisheries, on new modes of family relationships, etc. Here are some examples of StatCan publications that are of interest to the social sciences:

*   **Periodic publications:**

    Canadian Social Trends (quarterly)
    Canadian Economic Observer (monthly)
    Canadian Economic Observer: Historical Statistical Supplement (yearly)
    Canada Year Book (every second year)
    A Portrait of Canada (every second year, alternating with the Year Book)
    Survey Methods (twice a year)
    National Accounts (yearly)
    Distribution of Income in Canada (yearly)

    StatCan also publishes a daily report, but it is accessible only online (see below).

*   **Non-periodic publications**

    These are publications that examine in-depth one specific topic. Generally, all publications are available in both official languages. Here are some examples. The titles of the reports are followed by the catalog number of the publication.

    Time management (11-612 number 4)
    Elderly persons in Canada: Educational portfolio in Family Studies (11-519)
    Households and Environment (11-526)
    Adults with Disabilities: Their Employment and Education Characteristics (82-554)
    Lone-parent Families in Canada (89-522)

The general objective of such publications is to provide *interpretations* of the raw data. A general portrait of the situation is usually drawn, and data is given to present it or to illustrate it. Conclusions are often presented with supporting evidence. When frequency tables are given, it is usually to summarize a situation or to support a conclusion. Such reports are an interesting source of data, because they present it in ways that are concise and clear and they provide an interpretation of data that may be difficult to interpret by non-specialists. Specialists do refer to such analyses, but they also consult the raw data itself in order to come up with their own interpretation of the data.

Example

You want to find publications on family violence in Canada. If we look up the index of the 1994 catalog under the key word *Family*, we first find a suggestion to also consult the entries on three other keywords: *children, households*, and *persons alone*. We then find close to *five* pages of sub-entries! One of the lines under Families reads:

> Family violence, homicides, aggressions against the spouse and children, 89-523E

Going back to the first section of the catalog, you look for publication number 89-523E. You find it to be:

> **A portrait of Families in Canada/** Pina La Novara. 60 p. 28 cm.
> 1993. F and E.
> ISBN 0-660—94270-4.
> DDC 306.8/5/0971.

This is the full reference of the work. The F and the E that follow the year of publication mean that this publication is available in both French and English. If the library has this publication, you can locate it in the StatCan drawer numbered 89, as indicated in the publication number above. This is a detailed publication, as it contains 60 pages. You may want to look under other keywords to find more publications on the topic, and you may also want to consult the 1998 catalog as well as the comprehensive index of Canadian Social Trends, which contains shorter articles (4 to 6 pages). In the 1993–1997 index of Canadian Social Trends, under Family, you will find the following entry:

> Changes in Family Living   Summer 1993
>   *by Pina La Novara*

You may have noticed it is the same author who published the longer study mentioned above. The article in Canadian Social Trends is a concise article that focuses on one specific topic addressed by the more extensive publication mentioned above.

In that same index, under the subtitle **JUSTICE**, you will find the following entry:

Canada's Shelters for Abused Women   Autumn 1994
          by Karen Rodgers and Garry MacDonald

This has clearly to do with family violence, so you may want to retain that entry as well, as well as another one found further down:

Wife Assault in Canada   Summer 1994
By Karen Rodgers

If you now look in the 1986–1995 index of the same publication, you will find some of the same references, plus a few more:

Violence in the Family   Autumn 1989
  • Family Homicide
     by Holly Johnson and Peter Chislom
  • Male Violence in the Home
     by Eugen Lupri
Wife Abuse   Spring 1988
     By Holly Johnson

A more extensive search would include other keywords such as *Children and Violence*, for instance, or *family crimes*.

The preceding remarks suggest that before doing a search by keywords in the Index, you may want to draw a list of keywords to be used in the consultation, to make sure you look at all aspects of the issue under study. So, if your main keyword is family violence, you should include in your list of keywords the following:

Wife abuse
Battered women
Battered children
Elders
Children and violence
etc.

This list will be either short or extensive, depending on the intended scope of the research.

Now that you have determined a list of references to be consulted, you can read each one of them and dig out of it the information, data, and analyses that are relevant to your work.

CD-ROMS and Disks

StatCan makes available to libraries and researchers a number of CD-ROM disks that contain raw data. The most important is **E-Stat**, a CD-ROM disk that contains information taken from the latest censuses, from various surveys, and also **time series**. Time series are statistical information on a specific topic over many years, so as to allow the study of the evolution of a phenomenon over time. The **E-Stat** disk also contains data on the federal elections, indicating how many persons voted in each district, and which party they voted for. There are also disks containing environmental information and the list is constantly growing, as CD-ROMs increasingly become a usual research tool. In addition, a researcher who wants a specific statistical information that is not found on CD-ROMs can order it for a fee, and this information is then sent on a 3½ ″ floppy disk. In early 1999, StatCan made the **E-Stat** disk available directly online for educational institutions, which means that it can be accessed directly through the Internet. The organization of the material on the Internet is exactly the same as on the compact disc, except that the presentation (the graphics, the look) is slightly different.

Example

You want to find out how many men and how many women were elected under the banner of the various parties, in all the provinces, during the last election. When you open the **E-Stat** disk, you choose **Elections**. You are then given a choice between *Provincial Level* and *Federal Electoral District Level*. Choose *Provincial level*. You then have a choice among several profiles:

- Number of electors and polling stations
- Number of electors and polling stations for the 97, 93, 88, and 84 elections
- Number of ballots and voter turnout
- Voter turnout for the 97, 93, 88, and 84 elections
- Distribution of valid votes by voting method
- Distribution of valid votes under special voting rules
- Distribution of seats, by political affiliation and sex
- Number of valid votes, by political affiliation
- Percentage of valid votes, by political affiliation.

Let us choose one profile, say *Distribution of seats, by political affiliation and sex*. When you make that choice, you can choose the parties and the provinces for which you want to see such information. For this example, we will select the Liberal Party, the Bloc Québécois, the NDP, and the total for all parties. We will choose to see the number of seats that went to men and to women, in the provinces of Québec, Ontario, and Prince Edward Island (you can choose any combination of provinces and territories). We can then choose whether we want the display to be presented as a table, as a graph or as a map. The table display that is obtained from the *Internet version* of **E-Stat** is shown in Figure. 7.3.

Distribution of Seats, by Political Affiliation and Sex

Canada, by Province or Territory

Characteristic:	Ontario	Prince Edward Island	Quebec
Liberal Party of Canada - Male	76	4	20
Liberal Party of Canada - Female	25	0	6
Bloc Québécois - Male	0	0	33
Bloc Québécois - Female	0	0	11
New Democratic Party - Male	0	0	0
New Democratic Party - Female	0	0	0
Total - Male	78	4	57
Total - Female	25	0	18

⦿ No Sum   ⦿ No Sort
○ Sum starting with the [1st ▼] characteristic   ○ Sort on data for the [1st ▼] characteristic

                         ⦿ Actual Data
[Redisplay As:]   ○ Data as % of [1st ▼] characteristic

[Reduce/Sort Geo List]

[Reduce Characteristics List]

Figure 7.3 **Statistics from 1997 Federal General Election, by province or territory. (Source: E-Stat, Statistics Canada)**

The **E-Stat** disk also offers detailed tables taken from the 1996 Census as well as the 1991 census. They are organized on the same principle. The best way to become familiar with the information available is to spend time exploring the disk.

Online Data at StatCan

StatCan also makes available tables and reports on its website (http://www.stat can.ca). The home page of this website gives you a choice of French and English.

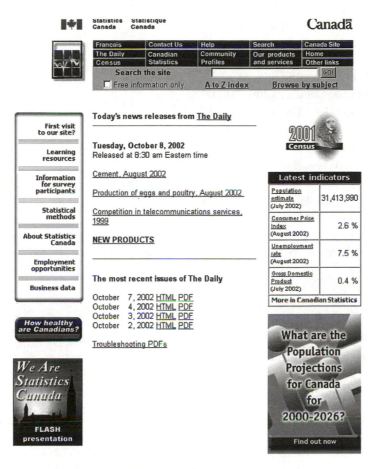

Figure 7.4 **Snapshot of the English home page of StatCan.**
(Source: Statistics Canada's Internet Site, http://www.statcan.ca)

When you click on the language of your choice, you access the main page, which is shown in its English version in Figure 7.4. Take a minute to look at the various links that are available. You can read daily reports and analyses, you can have access to the 1996 Census results, and you can also access a section on **Concepts, definitions and methods** where you can see in detail how the concepts are defined, how the samples were collected for the various surveys, etc. You can also consult or print a copy of the questionnaire used in a given survey, and finally you can have access to other electronic databases.

The Census is a rich source of data, and it is likely to be useful for background information when writing a paper on some social or economic issue in Canada. By exploring the various links you will realize that you can get information on about every social or economic aspect of Canadian life: population counts (even by municipality), statistics on education, work, income, family structure and number of children, etc. The **Nation Series** includes the following categories of tables (only one part of the list shown):

- Occupation, Industry and Class of Worker (20% Data)
- Household Activities (20% Data)
- Place of Work (20% Data)
- Mode of Transportation to Work (20% Data)
- Commuting Distance (20% Data)
- Education (20% Data)
- Mobility and Migration (20% Data)
- Sources of Income (20% Data)
- Family and Household Income (20% Data)
- Families (Part 2: Social and Economic Data) (20% Data)
- Occupied Private Dwellings (20% Data)
- Households and Housing Costs (20% Data)

Every one of these topics includes several tables that address various aspects of the topic.

Example

Find out the structure of families in Canada, that is, the number and percentage of husband-and-wife families versus single-parent families, and the number of children for the various categories.

   If we want to find analyses of the family situation in Canada, we should use the printed material of StatCan, as shown above. But if we just want to find the raw numbers, we can find them on the website of StatCan. Go to the website of StatCan, choose the language, then choose **Census**. Choose the link _The Nation Series- Selected Tables_. You get a new page displaying the following information:

1996 CENSUS: NATION TABLES

Families (Part 1: Number, Type and Structure) – (20% Data)

Census Families in Private Households by Family Structure, Showing Number of Families, Average Family Size and Number of Never-married Sons and Daughters at Home, for Canada, Provinces and Territories, 1991 and 1996 Censuses (20% Sample Data)

(HTML) (PDF)

Census Families in Private Households by Family Structure, Showing Counts and Percentage Change, for Canada, Provinces and Territories, 1986, 1991 and 1996 Censuses (20% Sample Data)

(HTML) (PDF)

Further data are available for Families in this product: Catalogue No. 93F0022XDB96008

Notice the following:

- The phrase **20% Data** appears several times. There are two questionnaires filled out in the Canadian censuses: a short questionnaire (A) given to every body, and a longer one (B) given to 20% of the population. The data used here is an estimate of the whole population generalized from the 20% sample.
- Every word that is underlined refers to a new page in the site; by clicking on it, you access that page.
- You can have access to two different tables, and the title of each is given.
- Each table comes in two formats: the HTML format allows you to see the table with the Internet browser you are using (such as Netscape, or Explorer, or any other browser), and the PDF format looks more like a printed page and requires another program (such as Adobe Acrobat) to be viewed. It is of course simpler to use the HTML format since it does not require opening another program.
- The two tables do not contain all the information available. You can get more detailed information by ordering it on a 3½ " floppy disk for a fee, using the catalog number provided.

If we now click on the HTML format of the first table, we get Table 7.2. We have reproduced here only the figures for Canada and have dropped the figures for the various provinces.

The table can now be read and interpreted as any other table. We can see that

Table 7.2   **Census families in private households by family structure, showing number of families, average family size and number of never-married sons and daughters at home, for Canada, provinces and territories, 1991 and 1996 Censuses – 20% Sample Data. (Figures for the various provinces/territories are not shown here)**

Family structure	1991 Census			1996 Census		
	Number of families	Average family size	Never-married sons and daughters at home	Number of families	Average family size	Never-married sons and daughters at home
Canada						
Total –						
Family structure	7,355,730	3.1	8,810,305	7,837,865	3.1	9,369,750
Husband-wife families	6,402,085	3.1	7,325,085	6,700,360	3.1	7,605,270
Now-married couples	5,682,810	3.2	6,841,705	5,779,720	3.2	6,869,700
Common-law couples	719,280	2.7	483,380	920,640	2.8	735,565
Lone-parent families	953,640	2.6	1,485,220	1,137,510	2.6	1,764,485
Male parent	165,245	2.5	248,425	192,275	2.4	276,710
Female parent	788,400	2.6	1,236,790	945,235	2.6	1,487,770

husband-and-wife families (whether married or common-law) added up to 6.7 million in 1996, representing 85.4% of families. Single-parent families represented 14.5% of all families (these percentages are not given in the table: they must be calculated). In 1991, single-parent families represented only 13% of the total of families, which is a relatively rapid change over a period of 5 years. The average family size given in the table includes the parents, so the average number of children for the husband-and-wife families is therefore $3.1 - 2 = 1.1$ (the average family size minus the two parents).

Finally, it should be pointed out that the main page of the StatCan website contains a link called *Links to other sites*. These links include three groups: Canadian sites, International sites, and Other sites. The international sites include many national agencies similar to StatCan in other countries. They can be visited to get information on other countries similar to the one found on the StatCan website.

Keywords

Statistical database	Primary data
Secondary data	Raw data
Index	Search by keyword
E-Stat	HTML format
PDF format	20% sample data
CSV format	GSS
ASCII format	

Suggestions for Further Reading

Internet for Social Statistics (http://www.sosig.ac.uk/vts/social-statistics/)

EXERCISES

Assignment on Accessing an Official National Statistics Database

7.1. Choose one of the national statistical sites available on the Internet and in your university of college library, and select one of the following topics:

Familyviolence   Aging in Canada   Families' net income

Crime in Canada   Juvenile offenders   Time-use
Leisure activities   Medical services   Educational level
Educational expenditures

(a) Find relevant statistical background information on the chosen website, and summarize it in at most one page.

(b) Conduct a search by keywords in the printed catalog of that institution to find an analysis of the situation, by going through the following steps:

    i   Write a list of 5 keywords for the bibliographic search.
    ii Find the entries for these keywords in the index or catalog of the institution. List the complete reference of 3 to 5 of the articles or reports that you have found.
    iii Select *one* article. Photocopy *the first page only* and annex it to your work.
    iv Read this article carefully, and write a one-page summary that includes:

        • the main idea (or conclusion) in the article (formulated in qualitative terms);
        • the main statistical facts that express this idea or that illustrate it;
        • the supporting evidence (more detailed statistics).

7.2. Perform a similar search on the Social Indicators site of the UN Statistics Division, using as a search term one of the following: poverty; water; sanitation; mortality.

8

STATISTICAL ASSOCIATION

The purpose of this chapter is to examine the basic meaning of statistical association with its important features (link, tendency, prediction, and strength), and then to see how statistical association is detected and measured depending on the level of measurement of the variables involved. The interpretation of statistical association as a qualitative relationship between the variables (explanation, possible causal factor, spurious association or other) is briefly discussed.

After studying this chapter, the student should know:

- the concept of statistical association and the fundamental aspects of a statistical association (link, tendency, prediction, strength);
- how to analyze association, depending on the measurement level of the variables;
- how to produce and read a two-way table (manually and with SPSS);
- how to produce and interpret a coefficient of correlation and a scatter plot;
- how to compare the mean of various subgroups on a variable;
- how to interpret a regression line, estimated scores, and errors in estimates;
- how to use the regression equation to predict a dependent variable;
- the difference between a statistical association and a relationship between variables;
- how to distinguish between the notions of explanation, causal factor, and spurious relationship.

The concept of statistical association is fundamental in research methodology. This concept allows us to formulate a clear notion of a *link* between variables when we notice that the scores of one individual on two different variables may somehow be related. But what do we mean by the word *related*? And how do we decide whether scores are related or not? Does it have to apply to every individual? Are there degrees in such relationships? What is the real meaning of statistical association? Does it mean that one factor is the cause of the other?

The notion of *statistical association* is quite abstract and it may be fuzzy for now, but we will gradually develop a detailed understanding of what it means.

Let us start with several examples.

- A teacher may notice that students who have good grades in mathematics tend to have good grades in physics as well.
- A doctor may notice that her female patients tend to be more resistant to certain kinds of infections than her male patients.
- A market study may demonstrate that people who like classical music tend to appreciate going to the opera more than those who do not like classical music do.

What do these statements exactly mean? Let us examine the first of our examples, which deals with the relationship between grades in mathematics and in physics. Suppose we have a class with the grades listed in Table 8.1.

Table 8.1

Student number	Grade in mathematics	Grade in physics
1	75	77
2	67	66
3	45	52
4	56	51
5	87	89
6	90	73
7	59	58
8	93	92
9	78	79
10	74	72
11	76	73
12	68	71
13	84	85
14	87	84
15	82	83
16	89	86
17	69	72
18	58	61
19	62	63
20	67	69
21	73	75

If we were to plot a scatter diagram of these grades in the two disciplines, we would get Figure 8.1.

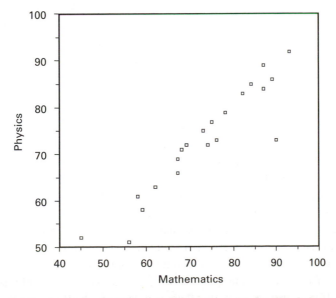

Figure 8.1 **Grades in mathematics and physics for a high school class**

Each dot represents one individual; the position of the dot with respect to the X-axis gives the grade of the individual in mathematics, and its position with respect to the Y-axis gives his or her grade in physics. Now we can identify several features in this diagram:

- When an individual scores low in mathematics, he/she tends to score low in physics as well.
- When he/she scores high in mathematics, he/she tends to score high in physics.
- Individuals whose score is close to the average in mathematics also tend to score close to the average in physics.
- The preceding remarks reflect a *tendency* and not a *rule*. You may have noticed that we always say that individuals who score in a certain way in mathematics *tend* to score in a certain way in physics. We can see that one individual does not fit the pattern outlined above, as this individual has a high grade in mathematics but a low grade in physics. This is why we talk about a *tendency* and not a *rule*.
- The notion of *prediction* is very important when we have a statistical association. If we know that somebody got a good grade in mathematics, we can *predict*, without knowing it, that his grade in physics is *likely* to be high. We see from the diagram above that we are right most of the time, but not all the time. Some individuals do not fit the pattern. This is why we use words like 'is likely to'. Predictions based on statistical association include a certain amount of error, in the sense that the predicted score differs from the real score by a certain amount, which is called the *error*. Such predictions also include a certain amount of risk, in the sense that there is a chance we are completely off track (as is the case if we tried to predict the grade in physics of the individual who got a good grade in math but a poor grade in physics).
- The notions of **dependent** and **independent variables** are used in this context. The *dependent variable* is what is to be explained, or what is to be predicted. The *independent variable* is the explanatory variable, or the variable used to make the prediction. In the example of the grades, the grade in mathematics is the independent variable and the grade in physics is the dependent variable. These two notions are not intrinsic to the variables, and the positions of dependent and independent variable could be interchanged, as we may want to see whether the grade in physics predicts the grade in mathematics with some accuracy.
- There are ways of measuring how *strong* an association is. The notion of *strength of an association* is related to that of prediction: if an association is strong, predictions based on it will tend to be good and will involve a small error. But if the association is weak, predictions based on it will often be way out … and involve large errors.
- The real concern here is to see whether there is some deep reason why people who perform well in mathematics also tend to perform well in physics. In some cases such a deep relationship exists, and in some others the statistical association is not indicative of a deep relation. Settling the issue of the existence of a relationship between variables is the real reason why we study statistical association. For the time being, let us remember that the existence of a statistical association is not a sufficient reason to say that there is a deep link between two variables.

The features outlined above express the essence of the notion of statistical association. But what if the variables are not quantitative? What does statistical association mean then? We will have to develop this notion separately for the various levels of measurements, and then draw some general conclusions. We will start by examining the case of two quantitative variables more closely.

The Case of Two Quantitative Variables

Let us suppose we have two quantitative variables, such as the grades of a class of students in mathematics and in physics in the example given above. We will denote the first one by X and the second one by Y. The grades of the various individuals in mathematics will be referred to as x_1, x_2, x_3, etc. and in physics as y_1, y_2, y_3, etc. When we want to talk about an individual in general, without saying which case this is, we will use the letter i. The situation is summarized in Table 8.2.

Table 8.2

Variable name	Symbol used	Entries are denoted by	General entry denoted by
Grade in Mathematics	X	x_1, x_2, x_3, etc	x_i
Grade in Physics	Y	y_1, y_2, y_3, etc.	y_i

Now we can start looking in more detail at the situation. Suppose the first student in the list has obtained 75 out of 100 in mathematics, and 77 out of 100 in physics, that is

$$x_1 = 75 \quad \text{and} \quad y_1 = 77.$$

This individual will be represented by the dot whose coordinates are (75, 77).

By looking at the scatter diagram shown in Figure 8.1, we can see a pattern. All the dots tend to fall on or near a straight line, called the **regression line**, shown in Figure 8.2.

This regression line represents the *trend* displayed by the dots. It can be described precisely by a mathematical equation (shown here at the top of the diagram). It can be used to **predict** the expected score in physics if the score of an individual in mathematics is known. On the diagram, you can see that somebody who scores 85 in mathematics is expected to score around 82 in physics: this is what the regression line suggests visually. If we want to calculate that predicted score more precisely, we could use the mathematical equation shown in the diagram, replacing x by the value 85. In this equation, y is the **predicted** value corresponding to a grade x in mathematics. This is what we get:

$$y = 11.523 + 0.83757x$$

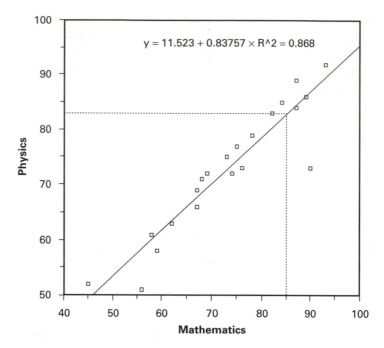

Figure 8.2   **Grades in mathematics and physics for a high school class**

If we replace *x* by 85 we get:

$$\text{predicted value of } y = 11.523 + 0.83757 \,(85) = 82.71$$

or 83 if we round up. You will notice that this is the predicted value. It is the *expected* score of the individual. Thus, the regression line and its equation allow us to predict the scores in physics of an individual whose score in mathematics is known. Some individuals' real score will be slightly above or slightly below the expected value. In one of the cases shown in Figure 8.2, the expected score will be *very* different from the real score: this is the case of the individual represented by the dot on the lower right of the diagram.

But how good are these predictions generally? Can we measure how good they are? The answer is Yes. To understand it, consider the situation of one individual, illustrated by Figure 8.3.

If the individual is far away from the regression line, using the regression line for prediction will yield a large error. But if the individual is close to the regression line, the error in predicting his or her *y*-score will be small.

When we consider the whole population from the point of view of prediction, we get six types of situations shown in Figure 8.4, diagrams (a) to (f).

In diagram (a), the points that form the scatter diagram and that represent individuals are all found to be close to the regression line. In this case, when the

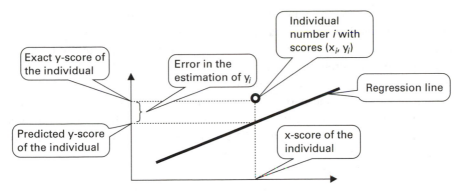

Figure 8.3

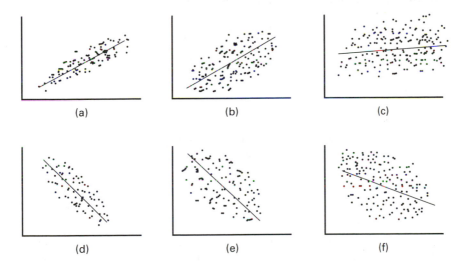

Figure 8.4

y-scores of individuals are predicted from their x-scores, the predictions tend to be generally good. We say in this case that **the correlation between the variable X and the variable Y is strong**. We used here the word *correlation* to refer to the statistical association. Indeed, correlation is the term to use when the variables are quantitative. Thus, the statistical association between quantitative variables is called a **correlation**.

In diagram (b), the points are not that close. We can still predict the y-score of an individual from his or her x-score, but the errors in prediction will tend to be larger than they were in diagram (a). In such a case, we say that the association between X and Y is not very strong.

In diagram (c), we see that the points are scattered far away from the regression line. People with high scores on the variable X do not tend to get high scores on Y: their scores on Y could be anywhere from low to high. In such cases, we say that **the correlation is weak** or even **null**.

The three remaining diagrams, (d), (e), and (f), are very similar to the preceding ones, with one difference that you may have noticed: as the x-scores increase, the y-scores tend to *decrease*. In such situations the **correlations are said to be negative**. They could be strong and negative, or weak and negative. The first correlations (a) to (c), in contrast, are said to be **positive**.

We have seen that some associations are weak (they yield poor predictions of the y-scores) and some are strong (they yield good predictions of the y-scores). In both cases they can be positive or negative. The next question now is to see **whether we can measure the strength of an association**.

There is indeed a mathematical formula that uses all the x- and y-values of the data to calculate the errors of prediction made on the basis of the regression line, and that comes up with a single number that summarizes it all. That number is called the **correlation coefficient**. It is obtained by the following formula.

$$r = \frac{\sum (x_i - \bar{x})\,(y_i - \bar{y})}{(n-1)\,s_x\,s_y}$$

where

x_i and y_i are the *i*th entry for X and Y respectively
$\bar{x}$ and $\bar{y}$ are the means of X and Y respectively, and
s_x and s_y are the standard deviations of X and Y respectively

This correlation coefficient is also referred to as the **Pearson product-moment correlation coefficient**. The values it produces range from −1 to +1. They can be interpreted as shown in Table 8.3.

To illustrate the use of the correlation coefficient, we can consider the numerical example given above. The diagram indicated that $r^2 = 0.868$, which corresponds to $r = 0.93$ approximately, and that is a very strong correlation. In SPSS, a simple command allows you to get the program to compute r and r^2 for any two numerical variables. You will learn how to do that in Lab 12.

Warning: SPSS will compute the correlation coefficient even when the variables are not quantitative, provided the codes are numerical values. In such cases, the correlation coefficient is not meaningful. You should use the correlation coefficient and interpret it only when the variables are quantitative and measured by a numerical scale. The correlation coefficient can sometimes be used for quantitative variables measured at the ordinal level, but its interpretation is trickier and these situations should be avoided at this stage.

The Case of Two Qualitative Variables

How do we know that there is a statistical association between variables measured at the nominal level? The method of the correlation coefficient shown above does not apply. To illustrate the situation, we will take a concrete example and analyze it.

Table 8.3 **Meanings of the various values of the correlation coefficient**

Value of r	Value of r^2	Meaning	Scatter diagram illustrating it
$r = 1$	$r^2 = 1$	The correlation is perfect and positive. All the points fall exactly on the regression line.	
$r = 0.8$	$r^2 = 0.64$	The correlation is positive and strong. The points are fairly close to the regression line and the predictions based on it tend to be good. *As r decreases, the correlation is still positive but weaker, the points tend to be scattered away from the regression line and the predictions are increasingly poor.*	
$r = 0.3$	$r^2 = 0.09$	Very weak positive correlation. Poor prediction of y on the basis of knowing x.	
$r = 0$	$r^2 = 0$	The correlation is null. Knowing the value of x does not tell us anything about the likely value of y.	
$r = -0.3$	$r^2 = 0.09$	Very weak negative correlation. Poor prediction of y on the basis of knowing x. *As r takes larger negative values, the negative correlation gets stronger, the points tend to be closer to the regression line and the predictions are increasingly better.*	
$r = -0.8$	$r^2 = 0.64$	The correlation is negative and strong. The points are fairly close to the regression line and the predictions based on it tend to be good.	
$r = -1$	$r^2 = 1$	The correlation is perfect and negative. All the points fall exactly on the regression line.	

In a survey conducted in a large company, 300 employees were asked whether they are socializing with their peers at work at a high level or at a low level, and whether they were planning to look for another job. Their answers were compiled in Table 8.4. Every rectangle in the table is called a **cell**. The numbers in the cells refer to the frequency of each category, and are called **observed frequencies**.

Table 8.4 **Cross-tabulation of the variables** *Level of socialization with peers* **and** *Intention to quit this job*

	Intention to continue with the present job	Intention to find another job soon	Totals
High level of socialization with peers	195	45	240
Low level of socialization with peers	40	20	60
Totals	235	65	300

A table such as Table 8.4 is called a **two-way table**, or a **contingency table**, or a **cross-tabulation** of the two variables. We can read in it that we have the answers for 300 employees, of which 240 have a high level of socialization with their peers, and 60 a low level of socialization. Of these same 300 people, 235 do not plan to leave their jobs for the time being, and 65 wish to find another job soon. The number written in the lower right corner is the *grand total*; the other totals are called *marginal totals*.

Can we determine, on the basis of that table, that there is some kind of link between the fact that people do not socialize with their peers and their desire to leave this job? In order to answer this question, it may be helpful to compute some percentages. We will compute the row percentages, that is, the percentages within the categories of socialization with peers. The results are shown in Table 8.5.

Table 8.5 **Cross-tabulation of the variables** *Level of socialization with peers* **and** *Intention to quit this job*

	Intention to continue with the present job	Intention to find another job soon	Totals
High level of socialization with peers	195	45	240
Percentage within Level of socialization with peers	*81.25%*	*18.75%*	*100%*
Low level of socialization with peers	40	20	60
Percentage within Level of socialization with peers	*66.6%*	*33.3%*	*100%*
Totals	235	65	300

We can now notice the following:

- Among those who have a high level of socialization with their peers, 18.75% plan to find another job. This is a little less than 1 person out of 5.
- Among those who *do not* have a high level of socialization with their peers, 33.3% plan to find another job. This is 1 person out of 3.

Remark. The question asked in the survey could be:

> Q. Would you say you have a high or low level of socialization with your peers at work? (check one)
>
> A. High level: _____
>   Low level: _____

But this is not a very good question, because there is no uniform definition of what is a high level or a low level. Instead, there could be a series of indicators, represented by questions such as:

> Do you take your lunch with your peers or alone?
> Do you walk or drive home with some of them?
> Do you phone some of them during the weekends?
> If you have a problem with the boss, would you trust them enough
>   to seek their advice?
> Etc.

On the basis of the answers to these questions, the researcher would divide the respondents into two groups: those who display a high level of socialization and those who don't. The criterion for classification could be something like: those who answered Yes to most of these questions will be classified as having a high level of interaction.

Here, the *concept* that we are trying to observe is the *level of socialization with peers*, and all the other variables (having lunch with them, calling them, etc.) are *indicators* of that concept. (Review Chapter 1 on these notions.)

We therefore notice a big difference between those who socialize with their peers and those who do not. In the latter category, a larger percentage of individuals plan to leave their job. We can say, therefore, that:

> **Individuals in this sample who do not socialize with their peers *are more likely* to want to find another job than those who do socialize with their peers.**

The preceding sentence illustrates the fundamental aspect of statistical association between two categorical variables: People who are in one of the categories of the first variable are *more likely* to find themselves in a given category of the second variable. Thus we can conclude:

> **There is a statistical association between the variables *Level of socialization with peers* and *Intention to quit this job*.**

Keep in mind, though, that it does not follow from that conclusion that the level of socialization is the *cause* of the intention to quit. It could well be the other way around. Or both variables could result from a third reason not presented in this table, such as: this place of work is in a remote area, far from people's houses. We will come back to the interpretation of the statistical association later in this chapter.

There is another way of looking at the statistical association described above. Instead of looking at the percentages within the levels of socialization, we could look at the percentage within the categories of the variable Intention to quit this job. We would get Table 8.6.

Table 8.6 **Cross-tabulation of the variables *Level of socialization with peers* and *Intention to quit this job***

	Intention to continue with the present job	Intention to find another job soon	Totals
High level of socialization with peers	195	45	240
Percentage within Intention to quit job	*83.0%*	*69.2%*	
Low level of socialization with peers	40	20	60
Percentage within Intention to quit job	*17.0%*	*30.8%*	
Totals	235	65	300
	100.0%	*100.0%*	

We can now make an analysis similar to the one we made above. Among the people who plan to continue working at the same place, 83% maintain a high level of socialization with their peers. But that percentage drops down to 69.2% among those who wish to find a job somewhere else. Thus, we can say that *the individuals of this sample who do plan to stay in this job tend to socialize with their peers at a higher level than those who plan to leave.* Again, this indicates (or confirms) that there is a statistical association between the two variables.

Note that the percentages written in the two tables above are called either:

> **row percentages** if they add up to 100% horizontally, across the cells of one row, or
> **column percentages** if they add up to 100% vertically, across the cells of one column.

You will learn in Lab 10 how to produce similar tables with SPSS. Keep in mind that we are only talking about statistical associations, not about causes. It does not follow from the existence of a statistical association that one of the variables is the cause of the other.

The Case of One Quantitative and One Qualitative Variable

Suppose now that we want to analyze the statistical relationship between one quantitative and one qualitative variable, for instance Income (quantitative) and Sex (qualitative). Several options are offered to us. The simplest is to compute the average of the quantitative variable separately for each category of the qualitative variable.

Example

The average income for a sample of 1500 people, consisting of 800 men and 600 women, is $19,400 a year. Suppose that the average income for women and for men separately is given by:

Average income of men: $23,400
Average income of women in that sample: $17,300
 This would mean that there is a large difference between the incomes of men and women. The income of men is (23,400–17,300) / 17,300 × 100 = 35.2% higher than that of women.
 This means that **there is a statistical association between the variables *income* and *sex*** for the individuals of that sample (we are not generalizing to the whole population yet). However, the preceding statement does not mean that sex is the *cause* of the difference in income. All we can say for the time being is that women make less money than men do. The *interpretation* of that difference is another matter. It could be due to discrimination (direct or systemic), it could be due to some other intervening variable (if, for instance, the women of this sample tended to be younger than the men, and therefore have less working experience) or some other cause.

 Finding the average for men and for women separately is not the only way to establish the existence of a statistical association. Another method would be to recode *income* into three categories: high, intermediate, low, and then treat both variables as categorical variables. In SPSS Lab 5, you have seen in detail how to illustrate the difference between the incomes of various groups graphically with box plots. SPSS Lab 11 shows how to compute statistical measures for each group separately.

Ordinal Variables

There are specific methods for establishing statistical association between ordinal variables. Such methods take into account the ranking of each individual on one of the variables in comparison to his or her ranking on the other variable. They will not be treated here. Ordinal variables are often treated as quantitative variables and correlations are computed. The results of such computations are sometimes difficult to interpret.

Statistical Association as a Qualitative Relationship

The interpretation of the statements made above in the section on two qualitative variables about the statistical association between them is not obvious. Recall that the two variables were the level of socialization of workers with their peers in a factory and their desire to stay or quit their job. We had found that the two variables were associated statistically. But there could be several possible interpretations of that statistical association.

First interpretation: We can interpret the statistical association to mean that a high level of socialization induces people to want to stay in that job. The explanation could be that the job is therefore more enjoyable, and people want to continue working there. In a way, the high level of socialization can be considered to be a cause for staying in that job, and inversely, a low level of socialization a reason to leave. So, we are now talking about more than a statistical association: we are talking about a **relationship** between variables. This situation can be represented by the diagram shown in Figure 8.5.

Figure 8.5

In symbolic terms, if we designate the level of socialization by X, and the desire to quit the job by Y, we could write:

$$X \Rightarrow Y$$

We could go a little further in that interpretation. If, in our theoretical framework, we had used the variable *Satisfaction with the job*, denoted by Z, as a general concept, and the level of socialization as one indicator of that concept, we could now conclude that the relationships can be illustrated by Figure 8.6.

Figure 8.6

The following pattern illustrates the situation.

$$X \Rightarrow Z \Rightarrow Y$$

In other words, the level of socialization is used as an **explanatory** variable, to explain why people are more inclined to quit their jobs. Notice that this interpretation does not follow from the statistical analysis of the association between the two variables. This is clearly an interpretation, and it is not the only possible interpretation, as we will see in what follows.

Second interpretation. We could reverse the preceding interpretation and say that if individuals tend to quit their job (they may perhaps want a better salary, or a more challenging job), they will not invest a lot of energy in socializing with their peers, since they know they are going to quit soon. Here the model is reversed:

$$Y \Rightarrow X$$

In other words, the desire to quit the job is used to explain why people do not socialize a lot with their peers. This interpretation, like the previous one, does not follow automatically from the statistical association between the two variables. The statistical association allows such an interpretation, but it does not prove it.

Third interpretation. The results of the statistical analysis are consistent with yet another interpretation, which asserts that both the desire to quit and the lack of socialization are the result of a third variable, such as *Desire to get a better salary*. If people think that their present salary is too low, and that they can get a better salary if they find another job, they may plan to quit and also they may decide not to invest too much energy and time in socializing with their peers. The model proposed here for explaining the statistical association is the following.

Fourth interpretation. The last interpretation that we could propose is to consider both variables as indicators of the general concept *Satisfaction with job*. This concept could be measured by several indicators: level of socialization, intention to stay, satisfaction with the salary level, pleasant atmosphere at the office, relationship of support and cooperation with the management, etc. In this interpretation, the key concept is the global satisfaction with the job. When people are globally satisfied, they are more likely to socialize with their peers, to consider staying in this job for a long time, etc.

Sometimes the qualitative relationship between two correlated variables is said to be *spurious*. To say that a relationship is **spurious** means that there is no logical link between the two variables, and that the statistical association is misleading. Such statistical association is often due to a third variable, but the logics linking each of the two correlated variable with the third one are completely unrelated. A classical example is that of height and salary. It could turn out that there is a statistical association between the height of an individual and his or her salary for a given sample. But if we break down the sample studied into men and women, we find that within each group there is no relationship. What happens is that on one hand men tend to be taller than women, and on the other hand in most societies the social structure favors men over women and the former end up tending to have higher salaries. The two kinds of associations (sex and height; gender and salary) follow logics that are totally unrelated to each other, hence our conclusion that the statistical association between height and salary is spurious. However, it is not always clear whether two sets of causal relationship are related or not, and one should be quite careful in interpreting a statistical association as spurious or as meaningful.

Summary and Conclusions

From Statistical Association to Relationship between Variables

The discussion above should help us understand better two distinct concepts, the concept of *statistical association* and the concept of *relationship between variables*.

Statistical association is something that can be observed objectively and measured, as we have seen in the examples above. Basically, it means that if you know the score of an individual on a variable X you can make a better guess of his or her score on another variable Y than if you did not know the score on X. The measure of statistical association depends on the level of measurement of the variables, which depends partly on the type of variables.

- For quantitative variables measured by a numerical scale, statistical association is called correlation. Two such quantitative variables are correlated when the values of one of them can be predicted with some precision from the values of the other variable. For linear correlation, the points representing the individuals are close to a straight line, which is called the regression line. If the association is strong, the points are very close to the line, the correlation coefficient r is close to 1 or −1, and the predictions based on the regression line involve a small error.
- For qualitative variables measured by a nominal scale, statistical association is analyzed with the help of a contingency table, also called a two-way table or a cross-tabulation. Statistical association means that individuals who are in a given category of the independent variable are more likely to be in a specific category of the dependent variable than in other categories. There are ways of measuring the strength of the association but they will not be discussed here.
- If one variable (X) is quantitative (measured by a numerical scale) and the other one (Y) qualitative (measured by a nominal scale), statistical association is studied by comparing the average scores on X across the various categories of Y.

This situation is summarized in Figure 8.7.

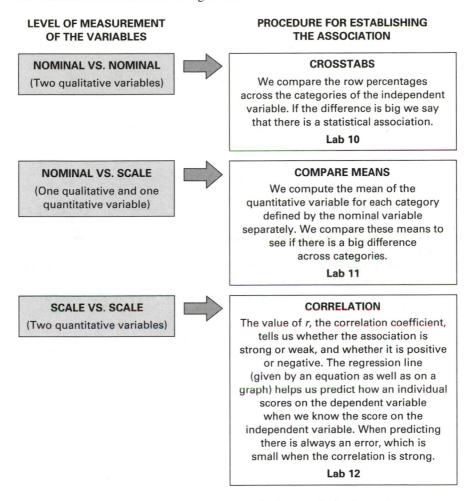

LEVEL OF MEASUREMENT OF THE VARIABLES	PROCEDURE FOR ESTABLISHING THE ASSOCIATION
NOMINAL VS. NOMINAL (Two qualitative variables)	**CROSSTABS** We compare the row percentages across the categories of the independent variable. If the difference is big we say that there is a statistical association. **Lab 10**
NOMINAL VS. SCALE (One qualitative and one quantitative variable)	**COMPARE MEANS** We compute the mean of the quantitative variable for each category defined by the nominal variable separately. We compare these means to see if there is a big difference across categories. **Lab 11**
SCALE VS. SCALE (Two quantitative variables)	**CORRELATION** The value of *r*, the correlation coefficient, tells us whether the association is strong or weak, and whether it is positive or negative. The regression line (given by an equation as well as on a graph) helps us predict how an individual scores on the dependent variable when we know the score on the independent variable. When predicting there is always an error, which is small when the correlation is strong. **Lab 12**

Figure 8.7   **How to measure statistical association? It depends on the level of measurement of the variable**

Relationship between variables. This notion is used to describe the *logical link* between variables. The independent variable could be a *cause* of the dependent variable, or an *explanatory factor* of the dependent variable; they could both be effects of some other variable; or they may be two indicators of a concept, or even two aspects of the same phenomenon. The notion of relationship between variables is a qualitative notion. It is a matter of interpretation, and it depends on the theoretical framework used in the research and on the research question or the research hypothesis. Statistical association should not be automatically interpreted as meaning a causal link.

Keywords

Statistical association	Prediction	Dependent variable
Independent variable	Strength of an association	Correlation
Coefficient of correlation	r	r^2
Positive correlation	Negative correlation	Regression line
Cross-tabulation	Contingency table	2-way table
Cell	Observed frequencies	Row percentages
Column percentages	Relationship between variables	

Suggestions for Further Reading

Berry, William Dale (1993) **Understanding Regression Assumptions** Newbury Park, CA: Sage Publications.

Fox, John (1997) **Applied Regression Analysis, Linear Models, and Related Methods**. Thousand Oaks, CA: Sage Publications.

Fox, John (2000) **Multiple and Generalized Nonparametric Regression**. Thousand Oaks, CA: Sage Publications.

Kenny, David A. (1979) **Correlation and Causality**. New York: John Wiley and Sons.

Rudas, Tamás (1998) **Odds Ratios in the Analysis of Contingency Tables**. Thousand Oaks, CA: Sage Publications.

Wonnacott, Thomas H. and Wonnacott, Ronald J. (1977) **Introductory Statistics** (3rd edn). New York: John Wiley and Sons.

EXERCISES

8.1   **Variables measured by a numerical scale (quantitative variables)**. The correlation coefficient between performance in math and in a language course among students in a college was found to be $r = -0.8$ (fictitious data). Draw a scatter plot that illustrates such a linear correlation and determine whether the following sentences are True or False (True if they correspond to such a correlation; False if they do not).

   (a)   The higher the grade a student gets in math, the lower he or she tends to get in a language course.

   (b)   Generally, as the grades in math increase, so do the grades in the language course.

   (c)   There is a strong association between competence in math and competence in language courses.

   (d)   High grades in language courses are associated with low grades in math.

   (e)   The cause of poor performance in math is high performance in language courses.

8.2 **Variables measured by a nominal scale (categorical variables)**. A study is conducted on the satisfaction of a sample of 530 people with their working conditions. The (fictitious) results are as follows:

	Men	Women	Row total
Very satisfied	180	35	.....
	.....	.....	.....
	.....	.....	.....
Moderately satisfied	120	25	.....
	.....	.....	.....
	.....	.....	.....
Not satisfied	100	70	.....
	.....	.....	.....
	.....	.....	.....
Column total	.....	.....	530
	.....	.....	

(a) Complete the following sentences:

  i Among the individuals in the sample, ...% are very satisfied with their work.

  ii Among the individuals in the sample, ...% are men.

  iii Among the men in the sample, ...% are moderately satisfied with their work.

(b) Indicate whether the following statements, based on the table, are TRUE or FALSE.

  i The majority of people who are very satisfied with their work are men.

  ii The majority of men are very satisfied with their work.

  iii There are more men who are *not* satisfied with their work than women.

  iv Men are more likely to be very satisfied with their work than women are.

(c) On the basis of the statements above, draw a conclusion: is there, for the people of that sample, a statistical relationship between Gender and Satisfaction with work? Give a reason for your conclusion.

8.3 **The regression line**. The following scatter plot illustrates the relationship between the total appraisal value of a house and its actual sale value in a municipality, measured in the local currency. The coefficient of correlation is given by $r = 0.787$, and the equation of the regression line is given by:

Sale price $= 14310.41 + 0.86 *$ (total appraisal value)

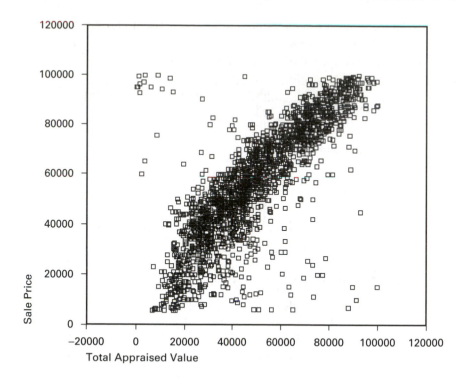

(a) Draw the regression line on the scatter gram.

(b) Find a house that fits the trend closely, indicate it on the scatter plot, and find its estimated sale price directly from the graph, then using the regression equation.

(c) For that house, give also its total appraisal value, and its actual sale price.

(d) Find the difference between the actual sale price and the estimated sale price (use the estimate from the regression equation).

(e) Repeat steps (b), (c), and (d) for a house that does not fit the trend.

9

INFERENTIAL STATISTICS: ESTIMATION

The purpose of this chapter is to explain the basic reasoning of inferential statistics, and then to show how confidence statements are to be made and interpreted. The calculations of the margins of error and the relationship between the confidence level and the margin of error are also shown.

After studying this chapter, the student should know:

- the meaning of inference in statistics;
- the notion of margin of error and probability of error;
- how to produce and interpret confidence statements involving means or proportions;
- how to determine the margin of error using either the table or the formulas;
- how to determine the size of the sample needed to achieve a certain precision;
- that the degree of precision increases with the risk of error.

Inferential Statistics

We have seen in the first chapter that there are two main branches of statistics, descriptive statistics and inferential statistics (refer to Figure 1.6). Chapter 3 was devoted to descriptive statistics. We are now going to study two main techniques used in inferential statistics, **estimation** (see Figure 9.1) and **hypothesis testing**, which are two distinct ways of drawing conclusions about a whole population when only a sample is known. This chapter will be devoted to estimation, and the next one to hypothesis testing.

Recall that **the purpose of inferential statistics is to draw conclusions about a whole population on the basis of information that has been collected on a sample**. In formulating such a generalization, we have to settle two issues that are closely related.

The first issue has to do with the **precision** of the results. Because the generalization is some kind of (educated) guess, it is never very precise. Therefore, we will have to introduce **a margin of error** in our statement, a term that will be defined

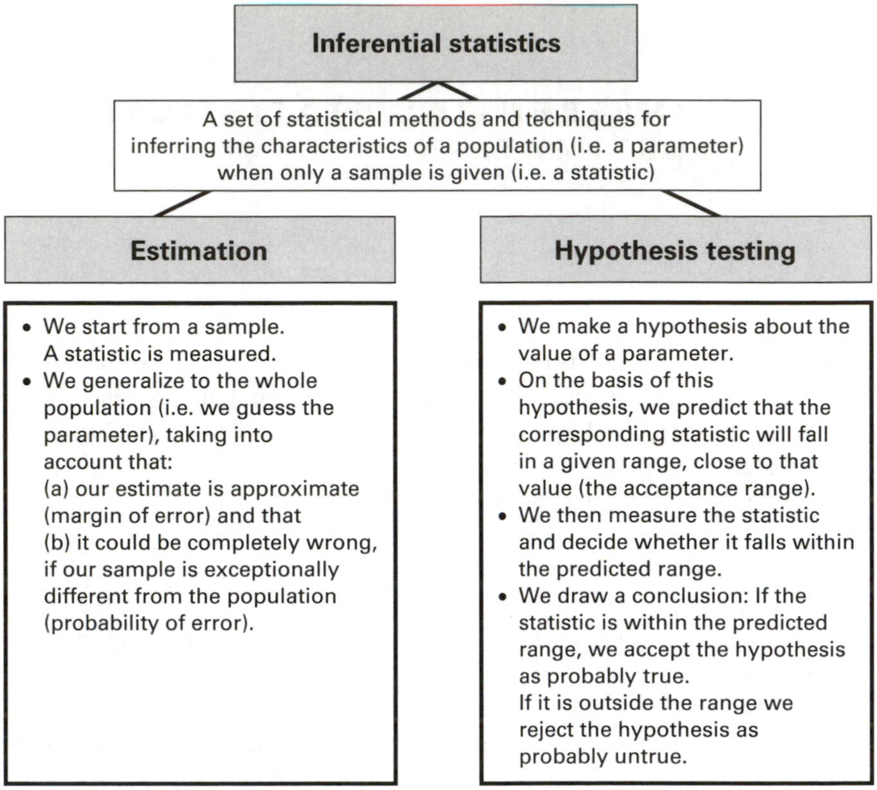

Figure 9.1   **Inferential statistics**

precisely below. For instance, if 45% of the sample of individuals who were interviewed answered Yes to some question, and if that sample of people is really representative, we estimate the percentage of people in the general population who would also answer Yes to be *around* 45%, not exactly 45%. May be somewhere between 44% and 46%, or between 43% and 47%. We will learn below how to determine this margin of error.

The second type of difficulty results from the randomness of the sample. We could be unlucky and hit a random sample that includes a large number of exceptional cases. Such a sample would not be representative, even if it had been selected at random. Only a small percentage of samples are likely to differ a lot from the general population, but the fact is that this possibility is very real. In order to take this possibility into account, we include in every inference a **probability of error**, which can be set at 10% or 5% or even 1%. Usually, the researcher sets out the risk he or she is willing to take when making a statement, and makes the inference on the basis of that level of risk. The precise way this is done will be explained below.

The Logic of Estimation: Proportions and Percentages

Suppose you select 200 students at random in your college, and ask them whether they approve or not a decision taken by the school administration about discipline in the college. Suppose also that 76% of them declared that they approved the decision. On that basis you are trying to guess what the percentage would be for the whole student population in your college, which comprises, let us say, 2400 students. What would you say?

You could say that 76% of the population approves of the decision. However, you can never be sure that this figure is accurate. It would be safer to say that you expect the corresponding percentage for the population to be *around* 76% rather than exactly 76%. You could say you expect it to be somewhere between 75 and 77%. Or somewhere between 74 and 78%.

The statement that results from your reasoning when doing an estimation is called a **confidence statement**. It is constructed as shown in the following example.

Example of a confidence statement:

> The poll, conducted on <u>1030 individuals</u> last week, showed that <u>37% of adult Canadians listen to the news on TV</u>. These results are accurate up to ± <u>4%</u>, and are reliable <u>95%</u> of the time. (fictitious data)

Let us examine the various elements that are included in the statement. They have been underlined, and they are explained below.

The population	The population here consists of all adult Canadians. Every confidence statement must specify clearly the population to which it applies.
The sample	The sample consists of 1030 individuals taken from the population. These are the ones that have been interviewed. On the basis of their answers, the results were extended to the whole population.
The variable measured	The variable measured here is whether the television is used as a source of information for news.
The measured percentage (the statistic)	The survey has shown that 37% of the people interviewed (that is, the sample) listen to the news on TV. This percentage was *measured* as part of a survey.

The estimated percentage (the parameter)	On the basis of that survey it is *estimated* that 37% ± 4% of the whole population listens to the news on TV. In other words, the estimation is that the percentage of people in the whole population that listens to the news on TV is somewhere between 33% and 41%, not exactly 37%. The middle point of that interval is 37% and this is called the **point estimate**.
The margin of error	The margin of error is ±4%. This is the degree to which the point estimate is accurate. When generalizing to a whole population, some accuracy is lost. The statement above says that the percentage of people getting their news from the TV is accurate up to + or − 4%. This is why the estimated percentage is not exactly 37% but somewhere between 33% and 41%. We will see below how this margin of error is calculated.
The level of confidence	The level of confidence here is 95%. It is a measure of how certain the results are. In other words, we are saying that 95% of the time, the sample we pick is sufficiently representative of the whole population to allow us to make a generalization. Details of that calculation will be discussed below.
The probability of error	This is the risk that the sample on which the estimation has been based was misleading and more different from the general population than expected. If the level of confidence is 95%, the risk is 5%. The level of confidence and the probability of error add up to 100%.

An important question has been left unanswered: How do we determine the margin of error and the probability of error?

The margin of error and the probability of error are closely linked. To explain this link, let us examine a familiar situation. It is a hot summer day. Two friends are arguing:

'I am sure it must be around 36° Centigrade, today. It is so hot!'
'Are you saying it is *exactly* 36°?'
'No. I'm saying it is probably *around* 36°. May be 35° or 37°. Something like that. I am almost sure.'

> 'Would you bet that your guess is correct and that the temperature is between 35° and 37°?'
>
> 'No. If you want to bet, I would say the temperature is between 34° and 38°. I am sure it must be within that range. I am ready to bet that it is within that range.'

What is going on in this discussion is that the first person is not ready to bet that the temperature is between 35° and 37°, and he figures out that there is a high risk of being wrong. However, he is more confident that the bet is correct when a wider margin of error is included. In that example, the risk of being wrong and the margin of error are not determined accurately. They are established on the basis of impressions. By contrast, in statistical inference, the level of confidence and the margin of error are determined precisely on the basis of a rigorous mathematical reasoning. However, the link between the two follows the same logic: if you want to make a guess with a high level of confidence, increase the margin of possible error. Give a wider range of possible answers: you will be more confident that the correct answer falls within that range. This relationship can be expressed in any of the following ways.

> To make estimations with a high level of confidence, we need to give a wide margin of error.
>
> Or: To diminish the probability of error, we need a wider margin of error.
>
> Or: In formulating an estimation, narrower margins of errors will necessarily imply higher probabilities of error.
>
> Or: Estimations that provide a wide range for the parameter can be done with a smaller risk of error than estimations that provide a narrower range.
>
> Or: Smaller margins of error are accompanied by greater risks of error.
>
> Or: Higher levels of confidence are accompanied by larger margins of error.

All these statements are logically equivalent and they express the relationship between the level of confidence and the margin of error in a confidence statement.

Estimation of a Percentage: The Calculations

The relationship between the level of confidence and the margin of error can be proven mathematically. Such a proof is beyond the scope of this text, but we can at least examine how it is *expressed* mathematically. Let us say that a survey involves a sample of size n, and that the proportion found in the sample is p. We can prove that in formulating a confidence statement about a proportion, the margins of error can be calculated with the formulas shown in Table 9.1.

Table 9.1 **Calculation of the margin of error**

If you want to be sure of your results at a 90% level of confidence	you must allow for a margin of error of $$\pm 1.64 \sqrt{\frac{p(1-p)}{n}}$$
If you want to be sure of your results at a 95% level of confidence	The margin of error is $$\pm 1.96 \sqrt{\frac{p(1-p)}{n}}$$
If you want to be sure of your results at a 99% level of confidence	The margin of error is $$\pm 2.58 \sqrt{\frac{p(1-p)}{n}}$$

Notice that the p used in the formula is a proportion, not a percentage. You can now verify that the statements made on the previous page are correct. Examine the various formulas carefully. They all look alike except for the coefficient that precedes the square root. As the level of confidence increases, the coefficient is higher, and it produces a wider margin of error.

Now look carefully at the numbers themselves. Do they ring a bell? Have we encountered these numbers before? You may recall that we have encountered them when studying normal distributions:

*   90% of all the data in a normal distribution falls within ± 1.64 standard deviation from the mean,
*   95% of all the data falls within ± 1.96 standard deviation from the mean, and
*   99% of all the data falls within ± 2.58 standard deviations from the mean.

For the 95% level of confidence, the margins of error corresponding to various sample sizes have been computed and presented in Table 9.2. It gives the approximate margins of error for various sample sizes and various values of the percentage. It can be used instead of the formula given above.

This is how you read the table: Suppose that in a survey of 539 people, it turns out that 62% of them answered Yes to some question. In the table, the closest column to 539 is the 500 column, and the closest percentage to 62% is the 'Near 60' percentage. The corresponding margin of error is underlined in the table: it is equal to $\pm 5\%$. What this means is that your estimate for the whole population will be $62\% \pm 5\%$, which is the same as saying it is somewhere between 57% and 67%.

NOTE: (for those who are not scared by a mathematical reasoning)

It is not a coincidence that these same figures show up again in this section. Indeed, suppose we had a population made of two subgroups A and B, with subgroup A forming a proportion p of the general population. If we formed all possible samples of size n taken from that population, and counted the proportion of people from group A in each of these samples, the set of all such proportions would constitute a distribution called the **sampling distribution**. We could prove the following: the sampling distribution is a normal distribution and its standard deviation, called the **standard error**, is equal to $\sqrt{\frac{p(1-p)}{n}}$. It follows that 95% of the values of that distribution fall within ± 1.96 standard deviations of that distribution of sample proportions (that is, standard errors). But these values are the sample proportions and each one refers to one sample of size n. This explains the figures in Table 9.1.

Table 9.2 **Margins of error for the estimation of a percentage, at the 95% confidence level.**

Population	Sample size						
Percentage	100	200	400	500	800	1000	1500
Near 10	7	5	4	3	3	3	2
Near 20	9	6	5	4	3	3	3
Near 30	10	7	5	5	4	3	3
Near 40	10	7	5	5	4	4	3
Near 50	10	7	5	5	4	4	3
Near 60	10	7	5	5	4	4	3
Near 70	10	7	5	5	4	3	3
Near 80	9	6	5	4	3	3	3
Near 90	7	5	4	3	3	3	2

Proportions and Percentages

The explanations given above apply equally to percentages and to proportions. The only difference is that a proportion is calculated out of 1 whereas a percentage is calculated out of 100. Thus, by multiplying a proportion by 100 we get the corresponding percentage, and by dividing a percentage by 100 we get the corresponding proportion. Some care must be given to the formulation of confidence statements in order not to confuse percentages and proportions. A given statement can be formulated either way. We could say, for instance, that an estimated percentage is 37% ± 4% or, equivalently, that the estimated proportion is 0.37 ± 0.04.

Point Estimates and Interval Estimates

You may have noticed that we have formulated the estimation in two different ways, one involving a *single value together with a margin of error*, and the other one in the form of a *range*. These two formulations are called, respectively, a point estimate and an interval estimate.

The **point estimate** in the preceding example is: 'The estimated percentage is 62% (± 5%).'

The **interval estimate** is: 'The estimated percentage is between 57% and 62%.'

These two formulations are equivalent and one can convert one into the other.

Formulation of the Level of Confidence

The level of confidence can be formulated as a percentage (for instance 95%) or as a ratio, as in 'These results are accurate 19 times out of 20.' The two formulations are equivalent, because if you multiply both numbers by 5 you get 'These results are accurate 95 times out of 100.'

For a level of confidence of 90%, the equivalent formulation would be: 'These results are accurate 9 times out of 10.' There is no similar simplification for a level of confidence of 99%.

Estimation of a Mean

The estimation of a mean follows exactly the same logic as that of percentage, except that the calculation of the margin of error is done with the help of a different formula. Here is an example.

> The poll, conducted on <u>1030 individuals</u> last week, showed that <u>adult Canadians</u> <u>watch the television</u> an <u>average of 4.3 hours</u> every day. These results are accurate up to <u>± 0.1 of an hour</u>, and are reliable <u>95%</u> of the time. (fictitious data)

Let us examine the various elements that are included in the statement. They have been underlined, and they are explained below.

The population	The population here consists of all adult Canadians.
The sample	The sample consists of 1030 individuals taken from the population. These are the ones that have been interviewed.
The variable measured	The variable measured here is the daily number of hours spent watching television.

The measured mean (**the statistic**)	The survey has shown that the people interviewed (that is, the sample) watch television for 4.3 hours (that is, 4 hours and 18 minutes) every day on the average. This average was *measured*.
The estimated mean (**the parameter**)	On the basis of that survey it is *estimated* that the population of adult Canadians spends on the average 4.3 hours daily watching television, with a margin of error of one-tenth of an hour (6 minutes). In other words, the estimation is that the **average** daily time people spent watching television is somewhere between 4 hours and 12 minutes and 4 hours and 24 minutes.
The margin of error	The margin of error is 6 minutes. We will see below how this margin of error is calculated.
The level of confidence	The level of confidence here is 95%.
The probability of error	The probability of error here is 5%.

The logic is exactly the same as in the case of the estimation of a percentage. Only the method for calculating the margin of error differs and we now turn to examining it.

Estimation of a Mean: The Calculations

Let us say that a survey involves a sample of size n, that the mean found in the sample is $\bar{x}$, and that the standard deviation for the population is σ. We can prove that in formulating a confidence statement for the mean of the population, the margin of error can be calculated as shown in Table 9.3.

Table 9.3   **Calculation of the margin of error when estimating a mean**

If you want to be sure of your results at a	you must allow for a margin of error of
90% level of confidence	$\pm 1.64 \dfrac{\sigma}{\sqrt{n}}$
95% level of confidence	The margin of error is $\pm 1.96 \dfrac{\sigma}{\sqrt{n}}$
99% level of confidence	The margin of error is $\pm 2.58 \dfrac{\sigma}{\sqrt{n}}$

There are no tables for the margin of error when estimating a mean, and these calculations must be done manually or with a calculator. SPSS computes the interval estimates, as explained in Lab 13. When the sample is large, the standard deviation calculated on the sample can be used instead of the standard deviation of the population.

For example, suppose a survey is conducted on a representative sample of 900 newborn babies in Canada and that it is found that their average weight at birth is 3.5 kg with a standard deviation of 0.5 kg. At the 95% level of confidence, the margin of error will be $\pm 1.96 \times 0.5 \div 30$ (which is the square root of 900), which gives approximately ± 0.033 kg, that is, ± 33 g (it is advised that you do the calculations yourself to make sure you understand the procedure). With this margin of error, we can come up with the following confidence statement:

The average weight of newborn babies in Canada is estimated to be 3.5 kg, with a margin of error of 33 g and a risk of error of 5%.

Or, equivalently:

At a confidence level of 95%, the average weight of newborn babies in Canada is estimated to be between 3.467 kg and 3.533 kg.

You may have noticed that the margin of error in this example is surprisingly small. This is because the sample is rather large. We are going to examine in some detail the effect of sample size on the margin of error.

As in the case of proportions, the estimate can be formulated as a *point estimate with a margin of error,* or as an *interval estimate* by subtracting and adding the margin of error to the point estimate, so as to get the whole range of values in which the estimated parameter falls.

Effect of the Sample Size on the Margin of Error

You may have noticed that, for both percentages and means, the formula giving the margin of error includes the root of n in its denominator, n being the sample size. If the sample size is 400, the formula includes 20 in the denominator. If the sample size is 900, the formula includes 30 in the denominator. This means that the margin of error gets smaller and smaller as the sample size gets bigger. In fact, we can make the margin of error as small as we wish by taking a big enough sample, but that may not be practical.

> The margin of errors gets smaller and smaller as the sample size gets bigger.

For instance, suppose that the standard deviation in the population is 12 units, and that you want a 95% level of confidence. A sample of size 100 would give you the following margin of error, calculated with the formula given on the previous page:

Margin of error for $n = 100$: $\pm 1.96 \dfrac{\sigma}{\sqrt{n}} = \pm 1.96 \times 12 \div 10 = 2.35$ units approximately.

If you want to improve your guess and make this margin of error *half as large*, you would have to take a sample *4 times bigger*. Indeed, if the sample size is 400 units instead of 100 units, you would be dividing by the root of 400, which is 20, and you would get:

Margin of error for $n = 400$: $\pm 1.96 \dfrac{\sigma}{\sqrt{n}} = \pm 1.96 \times 12 \div 20 = 1.18$ units approximately.

Conclusion: **In performing an estimation, every time you quadruple your sample size, you diminish your margin of error by one half**.

Or: **In order to make the margin of error half as large as the one we have obtained, we have to take a sample which is 4 times as big as the one we have.**

A similar calculation can be done for the estimation of a proportion, because the formula for the margin of error includes root n in the denominator. We can also conclude in this case that in order to cut the margin of error by half, we have to take a sample which is 4 times bigger.

Calculation of the Sample Size Needed in a Survey

The formulas seen above are useful for planning the data collection process in a survey. One of the steps of the design of a survey consists in determining the size of the sample needed. If we plan to make inferences about the whole population, and we want the margins of error to be reasonable, we have to select a sample that is large enough. But how large is large enough? If we make it larger than necessary, the survey might be more costly and longer than needed.

Examine Table 9.1, which gives the margins of error for the estimation of a percentage. You see that if your sample includes 100 individuals, you will get margins of error as high as 10%. Notice that for every sample size the largest margin of error corresponds to a percentage of 50%, which is the percentage you may find in a sample and that you wish to generalize. Suppose that you want a margin of error no greater than 4%. What is the sample size needed? Examining the table closely, you notice that by taking a sample of 800 individuals, the margins of error when generalizing will be 4% or less.

But we can also figure out the size of the sample needed to produce a given margin of error. To do this we have to isolate the n in the formula for the margin of error. For a confidence level of 95%, if m is the maximum margin of error you wish to allow, the sample size must be at least:

$$\text{Size of the sample } n = \left(\frac{1.96*0.5}{m}\right)^2$$

We used 0.5 instead of p in this formula because a proportion of 0.5 produces the greatest possible margin of error. If the p we are generalizing is other than 0.5, the margin of error will be smaller than the maximum we have set, which is fine. Keep in mind that the numbers must be entered in this formula as *proportions* (between 0 and 1), not as percentages. Thus if you want your margin of error to be at most 4%, you enter 0.04 as the maximum margin of error accepted. What the formula gives you is the size of the sample that will give you a margin of error equal or smaller than the maximum accepted. If you take a sample greater than the n you get from the formula, the margin of error will be even smaller.

A similar computation can be done when you want to generalize a mean. However, you must know the standard deviation of the population, or at least an estimate of it. If you reverse the formula given for the margin of error when estimating a mean, you get the following formula, where again m is the maximum margin of error allowed:

$$\text{Size of sample } n = \left(\frac{1.96*\sigma}{m}\right)^2$$

A sample of that size or larger will produce a margin of error smaller than or equal to the one we have set as the maximum margin of error allowed.

Summary and Conclusions

In this chapter we have seen how to estimate a mean or a proportion in a population when the corresponding statistic has been measured on a sample. In other words, we have estimated a parameter (mean or proportion) from our knowledge of the corresponding statistic.

Whenever an estimation is done, there is always a *margin of error* and a *probability of error*.

The **margin of error** reflects a lack of precision: the estimate is not exactly equal to the statistic, but falls *around* the value of the statistic, because every sample is likely to differ a little from the population.

The **probability of error** measures the risk that our estimate is wrong, that is, that the real parameter falls outside of the estimated range. This happens when the sample we have picked at random, and on which we base our estimate, differs from the population *more than expected*. The sentence 'differs from the population more than expected' means that the sample is an extreme case, presenting itself rarely. In an estimation, the risk of error that we are willing to tolerate is set first (usually at 1%, or 5%, or 10%), and then the margin of error is determined accordingly. When the risk of error is set at 5%, it means that 5% of all samples are considered to be extreme, or to differ from the population more than expected. Similarly, when the risk of error is set at 1%, it means that 1% of the samples are considered to be extreme, and when the risk of error is set at 10% it means that 10% of the samples are considered to be extreme. A notion complementary to the probability of error is the **level of confidence**, which is equal to 100% − (the risk of error).

As we said before, in an estimation we first choose the probability of error we are willing to allow (or equivalently the level of confidence we wish to have) and then we calculate the margin of error. This calculation is done with the help of the formulas given in the preceding sections. When estimating a proportion we could also use a table that gives the maximum margin of error that may result with a given sample size (Table 9.2).

The conclusion of an estimation is formulated as a **confidence statement**. The sections on estimating percentages and means have illustrated and explained all the elements that should appear in a well-formulated confidence statement. Finally, the estimation can be formulated either as a point estimate accompanied by a margin of error, or as an interval estimate that incorporates the margin of error within its range, as illustrated in Figure 9.2.

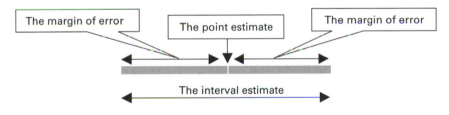

Figure 9.2

Keywords

Confidence statement   Point estimate
Interval estimate   Margin of error
Probability of error   Confidence level

Suggestions for Further Reading

Devore, Jay and Peck, Roxy (1997) *Statistics, the Exploration and Analysis of Data* (3rd edn). Belmont, Albany: Duxbury Press.
Wonnacott, Thomas H. and Wonnacott, Ronald J. (1977) *Introductory Statistics* (3rd edn). New York: John Wiley and Sons.
Wilcox, Rand (1996) *Statistics for the Social Sciences*. San Diego, CA: Academic Press.

EXERCISES

Interpretation of Confidence Statements

Analyze the following statements (numbers are fictitious) by providing the requested information.

9.1   A study conducted on 430 adult women in the Montreal area concluded that in general, when they have the possibility of either driving or taking public transportation, 73% of women tend to prefer driving. The results are accurate up to ± 5%, 19 times out of 20.

Variable studied: _____

Population studied: _____

Size of sample: _____ Statistic measured: _____

Estimated value of parameter: _____ Interval estimate: _____

Margin of error: _____

Probability of error: _____ Level of confidence: _____

9.2   On the basis of a survey conducted in your college, it can be asserted that on the average, students spend 43 minutes (± 11 minutes) each way commuting to and from school. There is a 10% chance that these results are wrong.

Variable studied: _____

Population studied: _____

Size of sample: _____ Statistic measured: _____

Estimated value of parameter: _____ Interval estimate: _____

Margin of error: _____

Probability of error: _____ Level of confidence: _____

Does this statement mean that each student spends somewhere between 32 minutes and 54 minutes in order to commute to school? _____ (Yes/No). Explain your answer.

9.3   In an internal poll conducted on the members of a party, the researchers concluded that party volunteers spend on the average 7 hours and 32 minutes every week working for the party. These results are accurate up to ± 45 minutes, 9 times out of 10.

Variable studied: _____

Population studied: _____

Size of sample: _____ Statistic measured: _____
Estimated value of parameter: _____ Interval estimate: _____
Margin of error: _____
Probability of error: _____ Level of confidence: _____

9.4 Canadians seem to prefer to spend their vacations home. Indeed, in a recent poll where 2045 people where interviewed over the phone, 69% indicated their intention to spend their vacation period in Canada. There is a 3% margin of error in this survey, and the results are reliable 19 times out of 20.

Variable studied: _____
Population studied: _____
Size of sample: _____ Statistic measured: _____
Estimated value of parameter: _____ Interval estimate: _____
Margin of error: _____
Probability of error: _____ Level of confidence: _____

9.5 The students of a college spend, on the average, somewhere between $2.45 and $3.15 in the cafeteria for their lunch. These results follow from 560 interviews conducted with students, and the results are trustworthy 9 times out of 10.

Variable studied: _____
Population studied: _____
Size of sample: _____ Statistic measured: _____
Estimated value of parameter: _____ Interval estimate: _____
Margin of error: _____
Probability of error: _____ Level of confidence: _____

9.6 Which of the following statements are equivalent to the statement of question 3 above?

(a) A party volunteer spends somewhere between 6 hours and 47 minutes, and 8 hours and 17 minutes every week volunteering in party activities. The level of confidence for that statement is 90%.

(b) The party's volunteers spend, on the average, somewhere between 6 hours and 47 minutes, and 8 hours and 17 minutes every week volunteering in party activities. The level of confidence for that statement is 90%.

Explain the difference between the two statements.

Formulation of Confidence Statements

In the following questions, you are given some information about surveys conducted on a representative sample. Complete the missing information and *formulate a confidence statement as a complete sentence.*

9.7   Variable studied: Smoking habits
      Population studied: All the employees of a very large company
      Size of sample: 238   Observed statistic: 29%
      Statistic studied: The percentage of respondents who smoke regularly
      Estimated value of parameter: _____   Interval estimate: _____
      Margin of error: _____
      Probability of error: _____   Level of confidence: 95%
      Statement:

      _____
      _____
      _____
      _____

9.8   Variable studied: Behaviour at stop signs
      Population studied: All drivers in a given city
      Statistic studied: The percentage of people who make a full stop
      Size of sample: 1200   Observed statistic: 89%
      Estimated value of parameter: _____   Interval estimate: _____
      Margin of error: _____
      Probability of error: _____   Level of confidence: 95%
      Statement:

      _____
      _____
      _____
      _____

9.9   Variable studied: Income generating work for full-time students
      Population studied: All college-level students in Australia
      Statistic studied: The average numbers of hours worked every week
                                   by respondents
      Size of sample: 900   Observed statistic: 15 hrs; st.dev = 3 hrs
      Estimated value of parameter: _____   Interval estimate: _____
      Margin of error: _____
      Probability of error: _____   Level of confidence: 95%
      Statement:

      _____
      _____
      _____
      _____

10

INFERENTIAL STATISTICS: HYPOTHESIS TESTING

The purpose of this chapter is to examine the logic of hypothesis testing and to apply it in simple cases of testing a hypothesis for a mean.

After studying this chapter, the student should know:

- the logical reasoning underlying hypothesis testing;
- the three forms of the alternative hypothesis;
- how to formulate the null and alternative hypotheses;
- how to determine the acceptance and the rejection regions;
- how to reach the decision to accept or reject an alternative hypothesis;
- how to perform and interpret simple t-tests.

Hypothesis testing is, with estimation, a main mode of reasoning in inferential statistics (see Figure 10.1). It is a process by which one proposition is accepted against another and which provides at the same time the probability of error in making that decision. To illustrate it, consider the following situation.

A research project has demonstrated that in a community college, let us say Dawson College, the percentage of students who say they get a high level of social support for their studies is 72%. Of course, such a result must be based on a clear definition of the concept 'social support', with explicit criteria for measuring all the components of social support (support at various levels: economic, psychological, technical, etc., and from a variety of sources: the family, the school, friends, etc.). You want to do a similar study on Champlain College, with the hope of demonstrating that the situation is different at Champlain. You will use the *same* questionnaire on a *sample* of Champlain students, but you must determine in advance how the results are to be interpreted. To do that, you formulate two hypotheses:

1. The **null hypothesis** states that the difference between the two populations is null, which means that we should expect that the same percentage of students in both colleges get a high level of social support. If the percentage of students who get a

Hypothesis Testing

The Logic ...   **and the Steps**

A hypothesis is formulated about the value of a parameter

This hypothesis is based on previous experience, or on an analogy between two situations. It assumes that the situation studied does not differ from the one we hypothesize, which is why it is called the

Null hypothesis, H_0.

An

Alternative hypothesis or research hypothesis, H_1

must also be formulated before we start the procedure: this is the hypothesis that will be retained if H_0 is rejected. For example, if we study the mean μ of a certain variable, we may have:

$H_0 : \mu = 34$		$H_0 : \mu = 34$		$H_0 : \mu = 34$
$H_1 : \mu \neq 34$	or	$H_1 : \mu < 34$	or	$H_1 : \mu > 34$

On that basis, a statistic is predicted
An acceptance region and a rejection region are determined

The reasoning is that if the sample is representative, it should not differ too much from the population. Some variability is allowed because the sample is not an exact replica of the population: there is a margin of error.

Thus, the prediction based on the first null hypothesis above is: the mean of the sample, $\bar{x}$, should fall between 32 and 36 (i.e. the margin of error is ± 2 units). This margin of error can be determined by the normal curve table, as we did for estimation, or using the t-distribution if the sample is small.

Acceptance region: $32 < \bar{x} < 36$
Rejection regions: $\bar{x} < 32$ or $\bar{x} > 36$

If the mean of the sample falls between 32 and 36, we accept the null hypothesis as probably true; otherwise, we reject it, considering that it is probably false, and we adopt the alternative hypothesis instead.

We now get a sample, we measure the variable, and we find the mean

The measurement can be obtained by a survey, or by an experiment, or by using existing databases.

Example: $\bar{x} = 35.7$

We make a decision

If the measured statistic falls within the acceptance region, we accept the null hypothesis. If it falls in the rejection region, we reject H_0 and accept H_1 instead.

In both cases, we run the risk of being wrong (this happens if the sample is not really representative of the population). Such risk can be calculated. The probability of being wrong is denoted by:

α for the probability of wrongly rejecting H_0

and

β for the probability of wrongly accepting H_0

Figure 10.1   **Hypothesis testing**

high level of social support at Champlain turns out to be equal to 72% or very close to it, you will conclude that Champlain students do not differ that much from Dawson College students. In other words, you accept the null hypothesis.

2. The **alternative hypothesis**, or **research hypothesis**, states that there is a real difference between the two populations for the variable under study. If you notice a big difference in the percentage of students who benefit from a high level of social support in the two colleges, you will conclude that the null hypothesis must be rejected, and that the alternative hypothesis (the one stating that the two populations do differ) is considered to be established.

This is, in a nutshell, the type of reasoning we are going to do. However, the explanations given above are not sufficient and they raise several unanswered questions.

First, how close is 'close'? If the percentage of students in your Champlain sample who get a high level of social support is 72.1%, is that considered 'close'? We could argue that after all, even if the sample is random, it is not an exact replica of the population and that you should expect some differences. The percentage in the student population at Champlain could well be 72%, but the sample could differ slightly, simply because it is just a sample, and include a percentage of 72.1%. Fine. We will accept 72.1% as 'close'. But how about 72.6%? How about 73%? Where is the cut-off point beyond which you say: no, this is too different from what I expected, the percentage in the population is probably different than 72%? The answer to this question will be given later in the chapter.

The second question has to do with the probability that we draw a wrong conclusion. What if the percentage in the student population at Champlain is indeed 72%, but that by a stroke of bad luck we hit a sample that is really different, giving us the impression that the whole population is different? Can we evaluate the risk of making such mistakes?

Fortunately, these two questions can be answered. We will give the answer and show why it makes sense. To do that, we must return to something just hinted at in Chapter 9 on estimation: sampling distributions.

Sampling Distributions

Suppose you are studying a variable X in a given population, and that the mean for X is given by μ_x. Suppose now you take a sample of some size, call it n, and you calculate the mean value of X for the individuals of that sample, which is denoted by $\bar{x}$. You can expect to get a variety of situations in the sample, some individuals scoring higher than the average μ_x and some scoring lower than it. If your sample is large enough (30 individuals or more) the values of X that are larger than the mean will more or less compensate for the values that are lower than the mean of the population as a whole, and the mean *of the sample* will end up being 'very close' to the mean of the population. By the phrase 'very close', we indicate that the means of the various samples end up being closer to the population mean than the individual observations. This property can be expressed mathematically in the following way. If we could take *all possible samples of size n*, and we calculated the mean of every one of them, we would get a distribution called **the sampling distribution of the mean**. If we draw a histogram for that distribution and compare it with the histogram of the whole population, we will see that the histogram of the sampling distribution is much *narrower* than that of the population, reflecting the property mentioned above, to the effect that sample means tend to be very close to the mean of the population. The standard deviation of the distribution of sample means is called the **standard error**, and it is denoted by $\sigma_{\bar{x}}$ which is read *sigma x bar*. We can prove mathematically that there is a relationship between the standard error (which is the standard deviation of the distribution of sample means) and the standard deviation of the population, given by the following equation:

$$\text{standard error: } \sigma_{\bar{x}} = \frac{\sigma}{\sqrt{n}}$$

This equation states that the dispersion of the distribution of sample means is much narrower than that of the population, by a factor of root *n*. If *n* is 36 (that is, if you take all samples containing 36 individuals) the square root of *n* will be equal to 6, which means that the curve representing the distribution is six times 'narrower' than that of the population. We can also prove mathematically that the sampling distribution is a *normal distribution* if *n* is large enough (*n* greater than 30). This means that all the properties of normal distributions apply. In particular, we can say that 95% of all sample means of samples of size *n* fall within 1.96 standard errors (recall that the standard error is the standard distribution of the distribution of all sample means). If the sample size is less than 30, an additional condition must be satisfied: the population distribution itself must be normal for the method to apply. If this condition of normality is satisfied, and if the standard deviation of the population σ is known, then the sampling distribution is normal. We will discuss below the situation where σ is not known.

> **Numerical example:** If the standard deviation of the population is 15 units, and you take samples of size $n = 100$, the standard error will be 1.5 units (15 divided by the square root of 100, which is 10). This means that the distribution of sample means for samples of size 100 has a standard deviation of 1.5 units. Therefore, 95% of all such sample means will fall within $\pm 1.96 \times 1.5$ units from the mean of the whole population, that is, 2.94 units.

The Logic of Hypothesis Testing

The calculations explained above will now allow us to specify what we mean by the sentence 'the sample mean is *close enough* to the predicted mean' and by 'there is a risk that our conclusions are wrong'.

To make the reasoning clear, suppose we are studying the variable **Height of Respondents**, and that the mean of that variable for the whole population is 172 cm, with a standard deviation σ equal to 5 cm. If we take all samples of size 100 (that is, containing 100 individuals) and if we calculate the mean of each one, they would form a normal distribution with mean = 172 cm, and with standard deviation = $5 \div \text{root } 100 = 5 \div 10 = 0.5$ cm. Thus the standard error is 0.5 cm. This means that the sampling distribution of the means can be represented by the symbol $N(172, 0.5)$ and that 95% of all samples of size 100 will have an average which is within $\pm 1.96 \times 0.5$ units from the mean of the population, that is, within ± 0.98 cm from the mean. In other words, 95% of all samples of size 100 have an average which is between 171.02 cm and 172.98 cm.

This calculation is the basis on which we will decide that a sample is 'close enough' to the mean of the population. If the sample means falls within the bounds we mentioned above, namely 171.02 cm and 172.98 cm, we will consider that this is close enough to the population mean. Why? Because 95% of the samples fulfill this condition. Only 5% of the samples will fall outside of these bounds. Therefore, 95% of the time, we are right when we say that the sample should fall within these bounds, and 5% of the time we are wrong. In conclusion we assert that:

> *a sample taken at random has a mean comprised between 171.02 cm and 172.98 cm. This assertion is true 95% of the time, and it is wrong 5% of the time.*

If we want to take a risk of being wrong smaller than 5%, say 1%, we have to adjust the calculations accordingly. If the risk of being wrong is 1%, this means that 99% of all samples should fall within the predicted range. The properties of the normal curve allow us to make such a calculation. Recall that in a normal curve, 99% of all entries fall within 2.58 standard deviations. If you apply this reasoning to the distribution of sample means with its standard error, you can conclude that 99% of all samples fall within $\pm 2.58 \times 0.5$ cm, which is equal to ± 1.17 cm approximately. The bounds within which 99% of sample means fall are therefore: 172 cm − 1.17 cm and 172 cm + 1.17 cm, or 170.83 cm and 173.17 cm. Therefore, we can assert that:

> *a sample taken at random has a mean comprised between 170.83 cm and 173.17 cm. This assertion is true 99% of the time, and it is wrong 1% of the time.*

This assertion is what explains and justifies the method of hypothesis testing. If you have a good reason to believe that the mean height of the respondents in a given population is 172 cm, and you want to test that hypothesis, you pick a sample at random from that population and you find the mean for that sample. If the mean is equal or close to 172 cm, you consider the hypothesis to be confirmed, otherwise you reject it. Now, we can give a very precise meaning to the sentence 'the mean is close to 172 cm', which depends on the risk of being wrong that you are willing to take. If you take a 5% risk, you look at 95% of the samples and you consider them 'close' to 172 cm. The calculations done above indicate that the samples we consider close have a mean between 171.02 cm and 172.98 cm. If the sample we have chosen at random has a mean that falls within these bounds, we consider that this is a confirmation (not a proof) that the mean for the whole population is indeed 172 cm. Otherwise, we conclude that the mean of the population is probably different.

This is the logic of hypothesis testing. We are now going to see the details of the procedure.

The Detailed Procedure for Hypothesis Testing

Let us now review the procedure from the very beginning. You start with two hypotheses about the mean of a population: one is called the null hypothesis and it states that the mean of the population is equal to some value c. The other is called the alternative hypothesis or research hypothesis and it states that the mean is different from c. You formulate them as:

Null hypothesis	H_0:	$\mu = c$
Alternative hypothesis	H_1:	$\mu \neq c$

There are other forms to the alternative hypothesis, and we will examine them later.

You must then determine the risk of error that you are willing to take, denoted by α, which is usually set either at 5% or at 1%. α is also called the **level of significance**.

On the basis of the risk of error, you will determine the **acceptance region** and the **rejection region**. The acceptance region is the range of values around the mean c that are considered to be close to c. The values that mark the boundaries of the acceptance region are called the **cut-off points**. Any sample mean falling beyond the cut-off points leads you to reject the null hypothesis.

With a risk of error of 5%, you take an acceptance region based on 95% of the samples as explained above, that is, $\pm 1.96 \times$ (the standard error). With a 1% risk of error you take $\pm 2.58 \times$ (the standard error).

You then compute empirically the sample mean of your sample, and you make the decision: if the sample mean is within the acceptance region, you accept the null hypothesis. If it is beyond the cut-off points, you reject the null hypothesis and accept instead the alternative hypothesis with a probability of error equal to α.

Numerical example

Test the hypothesis that the population of students at your college requires, on the average, 46 minutes to commute to school. The variable 'time to commute to school' has a standard deviation equal to 10 minutes. A random sample of

100 students answered a questionnaire to that effect and their average time to come to school is 47.2 minutes. Test the hypothesis at the 5% risk of error level.

Solution

Step 1: We formulate the two hypotheses:

Null hypothesis H_0: $\mu = 46$

Alternative hypothesis H_1: $\mu \neq 46$

Step 2: We formulate the acceptance region at the 5% risk of error level: General formula for the acceptance region:

$$c - 1.96 \frac{\sigma}{\sqrt{n}}, \ c + 1.96 \frac{\sigma}{\sqrt{n}}$$

Acceptance region in this case: $46 - 1.96 \times$ (standard error), $46 + 1.96 \times$ (standard error).

This is equal to: $46 - 1.96 \frac{\sigma}{\sqrt{n}}, \ 46 + 1.96 \frac{\sigma}{\sqrt{n}}$

After we replace σ by 10 and n by 100 we get:

$$(46 - 1.96, \ 46 + 1.96) = (44.04, 47.96)$$

The values 44.04 and 47.96 are the cut-off points. If the mean of the sample falls within these bounds, H_0 is accepted. If it falls beyond these values, H_1 is accepted with a probability of error equal to 0.05 (that is, a 5% risk of error).

Step 3: Interpretation. Since the mean of the sample is 47.2 minutes, it falls within the acceptance region. H_0 is accepted.
The situation can be represented by the diagram shown in Figure 10.2.

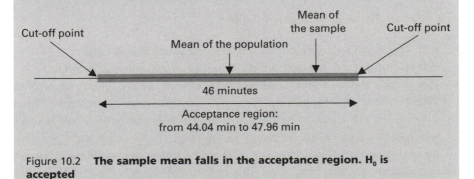

Figure 10.2   **The sample mean falls in the acceptance region. H_0 is accepted**

We could have a situation where the sample mean falls beyond the cut-off points. If in the previous example, the students in the random sample had indicated that they spend on the average 48.4 minutes to commute to schools, H_0 would have been rejected, a situation represented by Figure 10.3.

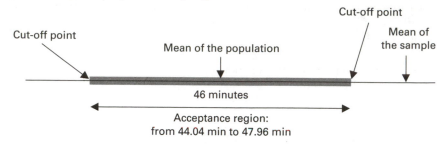

Figure 10.3   **The sample mean falls in the rejection region. H_0 is rejected and H_1 is accepted**

Understanding the Probabilities of Error

In the preceding explanations we have mentioned the probability of making one kind of error: rejecting H_0 when it should have been accepted. There is, however, another kind of error that could be made, which is accepting H_0 when it should have been rejected. The various possibilities can be represented by Table 10.1.

Table 10.1   **Errors of Type I and of Type II**

		In reality	
		H_0 is true	**H_1 is true**
Our conclusion	**H_0 is true**	Correct conclusion	Wrong conclusion. Error of Type II Probability of error: β
	H_1 is true	Wrong Conclusion Error of Type I Probability of error: α	Correct conclusion

We can see that there are four possibilities.

- If H_0 is true and we conclude that it is true, our conclusion is correct.
- If H_1 is true and we conclude that it is true, our conclusion is also correct.
- If H_0 is true and we reject it and accept H_1 instead, we are making an error of Type I. When could that happen? When the sample we have selected comes from a population with mean $= c$ as asserted by H_0, but falls among the 5% of samples which have a mean at the extremity of the range. In this case, the population mean is indeed equal to c, but a stroke of bad luck gave us a sample which is not representative of the population. This happens 5% of the time when we use the number 1.96 to calculate the acceptance region, and it happens only 1% of the time if we use 2.58 instead. This kind of error in called a **Type I error**, and the probability of making it is denoted by the symbol α. We usually determine α first (either 0.05 or 0.01), and on that basis we establish the acceptance region using

either 1.96 or 2.58 in the formula. The probability α is often called the **level of significance** of the test, because it indicates the proportion of samples that are considered to differ *significantly* from the mean of the population.

- If H_0 is false but we accept it by mistake, we call this a **Type II error**. Its probability is denoted by the symbol β. This type of error happens when the mean of the population is different from the one assumed in the null hypothesis, but the sample chosen happens to fall in the acceptance region, thus misleading us into concluding wrongly that H_0 is true. The probability β of making a Type II error cannot be computed easily, but we can say that if you use a small level of significance, you increase the probability β, and if you use a higher level of significance, you decrease the probability β. In other words, when α increases β decreases and vice versa, but the relationship is not inversely proportional. It is more complex than that.

A very important conclusion follows from the preceding discussion. That conclusion is to the effect that when we accept H_0 as true, we cannot determine with precision the probability of error. That probability could be rather large because the mean could be very close to c but not equal to it. On the other hand, the probability of making an error when accepting H_1 (and rejecting H_0) can be calculated precisely and in fact it can be determined by the researcher in advance. Therefore, when we assert that H_1 is accepted, such a conclusion is more solidly established for two reasons. First, we know precisely the risk of making an error, and second we can make that risk smaller, in fact as small as we wish, by taking a sufficiently large coefficient in the formula for the acceptance region (such a coefficient would be given by the table of areas under the normal curve). For these reasons, when you want to demonstrate that a hypothesis is true, the demonstration is more convincing, and it is built on more solid grounds if it is formulated as the *alternative hypothesis* of a test, not as the null hypothesis. This is why the alternative hypothesis is sometimes called the research hypothesis.

The Various Forms of the Alternative Hypothesis

In the preceding examples we have formulated the alternative hypothesis as $\mu \neq c$. There are, however, two other forms for it, in addition to the standard form. Here are the three forms:

First form:
$$H_0: \mu = c$$
$$H_1: \mu \neq c$$

Second form:
$$H_0: \mu = c$$
$$H_1: \mu > c$$

Third form:
$$H_0: \mu = c$$
$$H_1: \mu < c$$

A test based on the first form is called a **two-tailed test**, because the rejection regions are on both sides of the mean, forming the two tails of the distribution of sample means. The situation can represented by Figure 10.4 (recall that all sample means form a normal distribution if n is equal to or larger than 30).

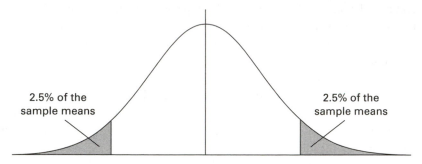

Figure 10.4

A test based on the second or third forms is called a **one-tailed test**, because there is only one rejection region forming a tail on one side only of the distribution of sample means.

The logic of performing a one-tailed test is very similar to that of two-tailed tests, except that the rejection region must all be on one side of the mean. When we determine the risk of making a Type I error, all the samples considered to be extreme or too far from the mean must fall on the same side of the mean. If we work with a 0.05 level of confidence, we must find a single cut-off point beyond which 5% of all sample means fall. This situation is represented by Figure 10.5 (recall that all sample means form a normal distribution when n is equal to or larger than 30).

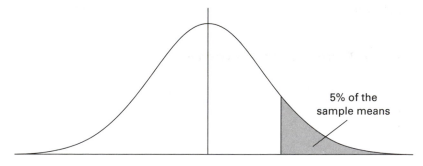

Figure 10.5

For a level of significance of 0.05 in a one-tailed test, the table of areas under the normal curves gives us 1.645 standard error. This is the coefficient to be used to determine the acceptance region, which will include all values equal to or less than $c + 1.645$ (the standard error). If rounding is desired, this number should be rounded up to 1.65 rather than rounded down to 1.64 to decrease the probability of error when accepting H_1.

When are One-tailed Tests Used?

One-tailed tests are used when we want to make a decision on whether a quantity is larger than a fixed value. Suppose for instance that you want to test if a teaching method increases the average grade in a class. You know the average grade for your population when the traditional method is used, and it is denoted by μ_{trad}. You conduct an experiment where the experimental group is taught with the new method. Since we are not interested in finding out simply if the new method produces a different average for the class, but if the new method gives higher grades, we establish the hypotheses as follows.

> **Null hypothesis H$_0$:** The average grade of the experimental group is *equal* to the average grade of the population when the traditional method is used.
> **Research hypothesis H$_1$:** The average grade of the experimental group is *higher* than the average grade of the population when the traditional method is used.

If we use symbols, we can reformulate these hypotheses as follows. Let $\bar{x}_{new}$ be the average grade of the class that was taught with the new method, μ_{new} the average grade of a population using the new method, and μ_{trad} the average grade of the population when taught with the traditional method. The hypotheses can be formulated as:

$$H_0: \quad \mu_{new} = \mu_{trad}$$
$$H_1: \quad \mu_{new} > \mu_{trad}$$

If we reject the null hypothesis when the level of significance is 0.05, let us say, we will conclude that the new method has the effect of raising the average grade, and the level of confidence in asserting this conclusion is 95%, which means that there is a probability of 5% that the sample chosen is extreme and that the increase in grade is due to chance and not to the method. In calculating the value of the cut-off point, we will have to use the formula:

$$\text{Cut-off point} = \mu_{trad} + 1.645 * \frac{\sigma}{\sqrt{n}}$$

where n is the number of individuals in the sample, and σ the standard deviation of the grades in the population taught with the traditional method. The research hypothesis will be considered established at the 0.05 level of significance if

$$\bar{X} > \mu_{trad} + 1.645 * \frac{\sigma}{\sqrt{n}}$$

Notice that we have used only one cut-off point, and we have used the value 1.645 and not 1.96, because we want the 5% area under the normal curve to be all on the right-hand side. If we accept the research hypothesis, we can do it knowing the risk of error that we are taking.

In this case, we assumed that the average grade of the population was known, and that its standard deviation was known also. It is often not the case. The following section examines variants of that situation.

Hypothesis Testing When σ is Unknown

In situations when σ is unknown, we can use the standard deviation of the sample, denoted by s, to estimate it.

If n is larger than 30, s is an unbiased estimator of σ and we can use it in the formulas explained above.

- Two-tailed test: Suppose the hypotheses are:

$$H_0: \quad \mu = c$$
$$H_1: \quad \mu \neq c$$

At the 0.05 significance level, the formulas for the cut-off points become:

$$c - 1.96 * \frac{s}{\sqrt{n}} \quad \text{and} \quad c + 1.96 * \frac{s}{\sqrt{n}}$$

- One-tailed test: Suppose the hypotheses are:

$$H_0: \quad \mu = c$$
$$H_1: \quad \mu > c$$

At the 0.05 significance level, the formula for the cut-off point becomes:

$$c + 1.645 * \frac{s}{\sqrt{n}}$$

Suppose the hypotheses are:

$$H_0: \mu = c$$
$$H_1: \mu < c$$

At the 0.05 significance level, the formula for the cut-off point becomes:

$$c - 1.645 * \frac{s}{\sqrt{n}}$$

But **if n is less than or equal to 30**, and σ is not known, we must use for the calculations the table of another distribution called the Student t-distribution (or simply a t-distribution). Recall that in this case the assumption of normality for the population itself must be valid. The table of the t-distribution is similar to the table used for the normal distribution, in that it gives the probabilities that a t-value will be greater than a certain quantity.

We can summarize the preceding discussion in Table 10.2.

Table 10.2 **Variants of the formulas**

Condition on sample size	Condition on population	Knowledge of population standard distribution σ	Conclusion
$n > 30$	None	Use σ, but s is accepted as an unbiased estimator of σ and it can be used.	Use the calculations and the table of the standard normal distribution.
$n \leq 30$	The distribution of the population must be normal	If σ is known,	... then you can use the table of the standard normal distribution.
$n \leq 30$	The distribution of the population must be normal	If σ is not known and you use s...	... then you must use the table of the t-distribution.

Hypothesis Testing with Two Independent Samples

Suppose you want to test whether two different samples, denoted by sample 1 and sample 2, come from populations that have the same mean or not. This means that you would be testing the hypothesis: $\mu_2 - \mu_1 = 0$. To test this hypothesis, you would compute $\bar{x}_2 - \bar{x}_1$ and see whether it is close to zero or rather far from it. The formulas for the cut-off points should take into account the fact that these two samples are independent, and that the standard deviations of the populations they come from are not necessarily equal. If the standard deviation for each population is known, and the sizes of the samples are denoted by n_1 and n_2 respectively, the formulas for the cut-off points at the 0.05 level of significance become:

$$(\mu_2 - \mu_1) - 1.96 \sqrt{\frac{\sigma_1^2}{n_1} + \frac{\sigma_2^2}{n_2}} \quad \text{and} \quad (\mu_2 - \mu_1) + 1.96 \sqrt{\frac{\sigma_1^2}{n_1} + \frac{\sigma_2^2}{n_2}}$$

Notice that this formula works if you want to test the possibility that $\mu_2 - \mu_1 = 0$, but it also works if you want to test the possibility that the difference of means of the two populations is a given fixed quantity other than zero.

Finally, if you are testing the difference in the means of two independent samples when you do not know the standard deviations of the populations, you use the standard deviations of the samples to estimate them as we did before, and the formulas for the cut-off points at the 0.05 level of significance become:

$$(\mu_2 - \mu_1) - 1.96 \sqrt{\frac{s_1^2}{n_1} + \frac{s_2^2}{n_2}} \quad \text{and} \quad (\mu_2 - \mu_1) + 1.96 \sqrt{\frac{s_1^2}{n_2} + \frac{s_2^2}{n_2}}$$

Of course, these formulas may be adjusted for the one-tailed tests as we did before: for the 0.05 significance level, we would use only one cut-off point and replace 1.96 by 1.645.

All these formulas can also be used for the 0.01 level of significance. In this case, 1.96 is replaced by 2.575 for the two-tailed tests, and 1.645 is replaced by 2.33 for the one-tailed test. These values all come from the table of areas under the normal curve.

Hypothesis Testing in Statistical Software

In most statistical packages, you do not have to determine the level of significance. The software will calculate the level of significance that would get you to reject H_0. The procedure works generally in the following way. The data in your document constitutes the sample. You enter the value of the mean to be used in the null hypothesis. The software evaluates the distance between the sample mean and the value you entered for the population mean, computes the probability p that a population with such a mean would produce such a sample, and labels it *Level of significance*. The level of significance p tells you the risk you are taking by asserting that H_1 should be accepted as true. If that number is very small (less than 0.05, or better less than 0.01) you can conclude that the null hypothesis should be rejected with a risk of being wrong equal to p. More details are provided in the lab on hypothesis testing.

t-Tests

The t-test is a procedure of hypothesis testing which involves small samples (less than or equal to 30 individuals). The procedure is very similar to the one explained above, the only difference being that instead of using the table of the normal distribution to determine the coefficients in the acceptance region, we use the table of the t-distribution mentioned earlier in the current chapter, which is calculated for small samples. The t-test can also be used for large samples, as the t-distribution looks increasingly like a normal distribution as n grows larger.

Summary and Conclusions

Hypothesis testing is one of the main modes of reasoning in inferential statistics. The diagram at the beginning of this chapter (Figure 10.1) summarizes the logic and the method.

In hypothesis testing you start with two hypotheses about the mean of a population: one is called the null hypothesis and states that the mean of the population is equal to some value c. The other is called the alternative hypothesis or research hypothesis and states that the mean is different from c. You must then determine the risk of error that you are willing to take, called the level of significance. It is denoted by α, and is usually set either at 5% or at 1%.

On the basis of the level of significance, you will determine the **acceptance region** and the **rejection region**.

At a level of significance of 5%, the acceptance region is bounded by the values

$$c - 1.96\,\frac{\sigma}{\sqrt{n}} \qquad \text{and} \qquad c + 1.96\,\frac{\sigma}{\sqrt{n}}.$$

At a level of confidence of 1%, it is bounded by the values

$$c - 2.58\,\frac{\sigma}{\sqrt{n}} \qquad \text{and} \qquad c + 2.58\,\frac{\sigma}{\sqrt{n}}.$$

These values, which mark the boundaries of the acceptance region, are called the **cut-off points**. Any sample mean falling beyond the cut-off points leads you to reject the null hypothesis and accept instead the alternative hypothesis, and this conclusion is reached with a probability of error equal to the level of significance.

There are two types of error in hypothesis testing:

- Type I error (rejecting the null hypothesis when it should have been accepted) occurs with a probability equal to α, the level of significance.
- Type II error (accepting the null hypothesis when it should have been rejected) occurs with a probability equal to β, which is difficult to evaluate but which increases when α decreases, and vice versa.

There are several variants to the hypothesis tests. We have mentioned the one-tailed and two-tailed tests, and the t-tests. Hypothesis testing can also be used to test whether two proportions are equal or different. The technical details vary, but the mode of reasoning is the same.

Keywords

Hypothesis testing	Null hypothesis	Alternative hypothesis
Research hypothesis	Sampling distribution	Standard error

Acceptance region   Rejection region   Cut-off points
Probability of error α, β Error of Type I (or Type I error)
Error of Type II   Level of significance   Two-tailed test
  (or Type II error)
One-tailed test

Suggestions for Further Reading

Devore, Jay and Peck, Roxy (1997) *Statistics, the Exploration and Analysis of Data* (3rd edn). Belmont, Albany: Duxbury Press.
Kanji, Gopal (1999) *100 Statistical Tests* (new edn.) London: SAGE Publications.
Wonnacott, Thomas H. and Wonnacott, Ronald J. (1977) **Introductory Statistics** (3rd edn). New York: John Wiley and Sons.

EXERCISES

Comprehension

10.1   Describe briefly the main steps in hypothesis testing.

10.2   In a survey on poverty and violence, a researcher wants to establish beyond any doubt that there is causal relationship between poverty and violence within the family. Is the method of hypothesis testing appropriate for that purpose? Is any other method appropriate for that purpose?

10.3   Indicate which, in the following formulations of the null and alternative hypotheses, are correct and which are wrong for the purposes of hypothesis testing.

(a) H_0: $\mu = 20$
      H_1: $\mu < 30$
(b) H_0: $\mu = 30$
      H_1: $\mu < 30$
(c) H_0: $\mu = 20$
      H_1: $\mu > 30$
(d) H_0: $\mu \neq 20$
      H_1: $\mu = 20$
(e) H_0: $\mu = 30$
      H_1: $\mu < 30$
(f) H_0: $\mu \leq 20$
      H_1: $\mu > 20$

10.4   Formulate a null and an alternative hypothesis for each of the following situations.

(a) You wish to test the hypothesis that the response time of a group of subjects to a psychological test is different than that of the general population, which has been established as 34 seconds.

(b) You wish to test that the response time in (a) is larger than that of the general population.

(c) You wish to test the hypothesis that establishing a day-care center at the place of work diminishes the number of days off per year taken by female employees who have young children. The average rate for this category of employees has been established to be 5.2 days a year when there is no day-care.

(d) The hypothesis to be tested is that the rate of children affected by a certain disease diminishes when a vaccine in administered in schools. The rate when the vaccine is not given is 22 cases per 10, 000 children a year.

Applications

10.5 In each of the following situations, formulate in plain language:

- the objective of the use of the method of hypothesis testing,
- the conclusion that is reached given the information provided, and
- determine if the error that would be made, if the conclusion was false, would be of Type I or Type II.

(a) Sample: the graduating students of a secondary school in a country. Population: the graduating students of the whole country. Variable: the average grade obtained by the students in their last year of secondary studies, denoted by g_s, and that of the students of the whole country, denoted by g_c.

$$H_0: g_s - g_c = 0$$
$$H_1: g_s - g_c > 0$$

The value $g_s - g_c$ turned out to fall in the acceptance region.

(b) Same as in part (a), but the value $g_s - g_c$ falls now in the rejection region.

(c) Sample: all students in a college. Variable: the social support (denoted by s) they get in their studies, measured on a Likert scale that ranges from 1 to 4. The social support their predecessor got in the previous year averaged 2.9 on that scale.

$$H_0: s = 2.9$$
$$H_1: s \neq 2.9$$

For that sample, the average measured turn out to fall in the acceptance region.

10.6   We want to test the hypothesis that a new method of treatment allows patients suffering from a given disease to be released from hospital faster than with the traditional cure. The sample contains 36 patients. The average length of hospitalization with the traditional method is 18 days, with a standard deviation of 3 days.

   (a)   Determine the acceptance region and the cut-off points at the .95 level of confidence.

   (b)   If the average length of stay for the patients following the new cure turns out to be 16.5 days, what is your conclusion? Are you risking making an error of Type I or of Type II?

   (c)   Answer the questions of part (b) if the average length of stay was instead 17.3 days.

SYNTHESIS: ANALYZING A TOPIC IN A DATA FILE

A reminder of the main issues that have to be addressed when analyzing a data file

The purpose of this synthesis is to recapitulate the process to be followed when performing the analysis of a data file. It elaborates the explanations given in Chapter 4, and expands them to include statistical associations and inferential statistics, and to include as well what has been learned in the labs. It integrates what has been learned in the various chapters in this book. It presents a framework that can be expanded to accommodate more advanced statistical methods.

The Data File Itself

Give a brief description of the data file. Where does the data come from? When was it collected? By whom? With what purpose? Was it a survey? How many cases are involved? What was the sampling method used? Did it come from a statistical archive? Etc.

You may not have the answers to all these questions, specially in the context of a course, but ideally, these pieces of information would be useful.

The Topic and the Questions

You must start by clarifying what is it you want to analyze in the data file. Do you have a specific research question? Or are you simply summarizing what is known about a specific topic (for instance 'education' in a General Social Survey)? At that point, you may want to specify the dependent variables (the variables you want to study or explain) and the independent variables (the possible explanations).

For example, the dependent variables could be a specific kind of behavior, of attitudes, or of opinions or beliefs. The independent variables could be socio-economic variables, which are indicators of the position of the individual in society (age, sex, level of education, income, marital status, type of job), and which may explain one or more of the dependent variables.

Examples

1. Is political tendency (a dependent variable) explained, at least partly, by educational level (an independent variable here)? Is it explained by income? By sex? By religious beliefs? By other independent variables? If more advanced methods are used, we can find which combination of these variable best describes the dependent variable.
2. Is consumer behavior such as going to the restaurant, buying certain brands of clothing (the dependent variables), explained by these socio-economic variables (which are treated as independent variables)?
3. The socio-economic variables themselves may also become dependent variables and be explained. We could ask whether the educational level of individuals (a dependent variable now) can be explained by the income of their parents (the independent variable); or by their cultural values (another independent variable); or by a combination of these and other variables as well.

The Analysis

Univariate or Bivariate?

Independent variables are not analyzed the same way as dependent variables. For independent variables, you just want to summarize each variable by itself, in order to give a portrait of the sample or the population you are analyzing. For instance, you may want to say what is the percentage of men and women, the average age of the group, etc. This is a univariate analysis.

For dependent variables, you may want first to give a univariate description of each, and then find statistical associations between these dependent variables and some of the independent variables. This process constitutes a bivariate analysis.

For example, if your focus is political views or party identification, you first want to summarize the data of the variable itself, as explained below. Then, you want to see whether gender matters, or income, or education. In other words you want to find statistical associations between the dependent variables of political view and party identification on one hand, and the independent variables: sex, income, and education on the other hand.

You will have to decide whether a given variable deserves simply a univariate description, or whether you want to compare it with other variables, or split your population into groups to see how each group scores on that variable.

For example, you may want to split your population into men and women and see how each group scores on a certain variable, even for background socio-economic variables such as educational level or income.

Dependent Variable Measured at the Nominal Level

DESCRIPTION

Do a frequency distribution to find out the percentage of units in each category. Then ask yourself these questions: Are there lots of missing values? In this case, there is a difference between the total percentage and the valid percentage of each category. Which is more relevant for that variable, the total percentage or the valid percentage? Are the categories completely distinct, or is there a reason to group some of them together?

Recall that in one of the files we have studied, the variable **Marital status** had the following categories: marriage only, common law then marriage, common law only. After giving the frequency table that includes the three categories, it makes sense to group together the first two: these are the people who are presently married, and to indicate what percentage of people lived together in common law before getting married.

GRAPHICAL REPRESENTATION

As for graphical representations, you have a choice between pie charts and bar charts. Usually, you use a pie chart when the emphasis is on the *distribution* of a fixed quantity over the various categories (that is, the percentages). Bar charts are better when the emphasis is on the *amount* of units in each category (that is, the count).

Example

If the dependent variable is budget items, and the budget is the family budget, you may want to emphasize the fact that low-income families tend to spend more than a quarter of their income on housing, whereas middle-class and upper-class families tend to spend less than 25% of their income on housing. In this case, you would use a pie chart showing the various categories of spending (housing, clothing, education, food, etc.) for low-income families, and another pie chart for medium- and high-income families. But if the focus were on the actual cost of housing and not the proportion of budget spent on housing, you would use a bar chart.

STATISTICAL ASSOCIATION

If you want to see whether various groups of people tend to fit into different categories of that variable, you will have to do a cross-tabulation (under the **Summarize** sub-menu in SPSS). If you want to see whether people who have a higher score on a certain ordinal or numerical scale variable differ from those who have a lower score, you must use a **Compare means** procedure.

Example taken from the **GSS93 subset** file in SPSS. Dependent variable: Party Identification, that is, **partyid**. Independent variables: Respondent's Sex, and one of the income variables (there are three variables measuring income). First, to simplify the analysis, we want to recode Party identification (partyid) into a new variable, **partyid3**, having three categories: Democrat, Independent, and Republican. Then we can do a cross-tab between **partyid3** and **sex** to see whether women and men tend to differ in their party identifications. To study the link between **partyid3** and income we have a choice. We can use Respondent's income or Total family income, and then we can do a Compare Means procedure. We will have the average income of three separate groups: those who are Democrats, or Independent, or Republican. We can then examine the tables to see whether the differences between the three groups are big. We can also do a cross-tab procedure between **income4** and **partyid3**, since these two variables have a small number of categories. Try both procedures to see the difference in output: the conclusion will then become clear.

CONFIDENCE STATEMENTS

If you want to generalize the findings to the whole population (that is, formulate a confidence statement), you must start from the frequencies given in the table, and then add a margin of error using either the formulas given in Chapter 9 (p. 166), or the table on p. 167 that gives the approximate margins of error at the 95% confidence level (for the 90% and 99% confidence levels, or if the sample size is not close to the ones given in the table, you must use the formulas).

Dependent Variable Measured at the Ratio/Interval Scale Level

DESCRIPTION AND GRAPHICAL REPRESENTATION

A quantitative variable is best summarized by its mean, median, and mode, which give an idea of the central tendency. Recall that the relationship between the mean and median is important: if they differ a lot, it means that the distribution is skewed. In this case, the median tends to be a better representative of the bulk of the data. You must also give some information about the dispersion of the data. This is done by giving the minimum and maximum values, the 1st and 3rd quartiles, and of course the standard deviation. The histogram is usually the most appropriate type of graph.

When the values have been grouped, it can become tricky, because the mean is not the real mean of the variable, but the mean of the *codes* used to denote the various categories. So it could be a little misleading, specially if the categories do not all have the same range (or length), or if there is a category that includes all values larger than a certain value. Still, the mean is useful for making comparisons (recall the example of the Respondent's income in the GSS93 file). In all these cases, the level of measurement has become in reality ordinal. When there are few categories (let us say 5 or so) the relative frequencies of the various categories become more important than the mean, and a bar chart becomes more appropriate. For a variable where you have a few discrete values that have not been grouped, both the mean and the relative frequencies are important.

Example

Consider the variable Number of children. It is a quantitative variable, but the actual values are just a few. Most families in Canada, for instance, have fewer than 5 children. This variable is best described by the average number of children, but the percentage of families with one kid, or two kids, etc. given by the frequency table, is also a useful description of the situation. In that case, the percentage of families with 6 kids or more is very small, and you can lump them together.

STATISTICAL ASSOCIATION

If the independent variable is also quantitative, and measured by an interval/ratio scale, you can compute the coefficient of correlation and graph the scatter plot. But if the independent variable is measured at the ordinal or nominal level and if it has a small number of categories, the **Compare Means** procedure is more appropriate.

CONFIDENCE STATEMENTS

Here, you can simply use the **Explore** command, which gives you the confidence interval directly, at the confidence level that you choose. You can then compute the margin of error yourself by a simple subtraction: the upper level of the confidence interval minus the mean.

Dependent Variable Measured at the Ordinal Level

DESCRIPTION AND GRAPHICAL REPRESENTATION

These variables share properties of both nominal scales and interval/ratio scales. They usually comprise a small number of categories, and in this case they can be analyzed like variables measured at the nominal level of measurement. You must therefore give the relative frequency (or percentage) of each category, and use bar charts or pie charts. For nominal variables, you are more likely to want to group categories.

> ### Example
>
> The question is: Do you Strongly Agree, Agree, Disagree, or Strongly Disagree with the idea X? This is an ordinal scale with four levels (recall that scales where people are asked to rank their opinions from negative to positive, with 3, 4, 5 or more categories, are often called Likert scales). It makes sense to give the table of frequencies, for all four categories, but in the explanatory paragraph that accompanies the table, group the Strongly Agree and the Agree together, and the Strongly Disagree and the Disagree together.

In addition, you may use procedures that are more specific to the numerical scale level. For instance, it often makes sense to find the average. If people are ranking a service or a product from 1 to 5 (1 = very bad; 5 = very good), you can find the average score for your sample.

STATISTICAL ASSOCIATION

There are procedures for measuring statistical association that are specific to the ordinal level of measurement. They were not included in this book, but we may be able to use cross-tabulation, or the Compare means procedure, or correlation. The procedure used will depend on the number of categories. If the ordinal variable has a large number of categories (let us say more than 10), it may make sense to find the coefficient of correlation, but the scatter plot will not be useful, as a variable number of points would appear as a single dot in the diagram. If it has fewer than five categories, you can use either the Compare means procedure (if the independent variable is quantitative) or the cross-tabs procedure if the independent variable is grouped into categories (nominal or ordinal).

> ### Example
>
> If you have the scores people gave to evaluate the quality of the service in a restaurant, you can see whether people who were accompanied by no children, one child, or two children or more tended to give different ratings to that restaurant. Such a comparison makes sense.

CONFIDENCE STATEMENTS

Both procedures (those used for qualitative variables and those used for quantitative variables) can be useful. You must judge which is more relevant depending on the number of categories used.

Finally, recall that a descriptive report must summarize the essential numerical features of the data in plain, direct language, and use the tables as supporting evidence. The diagrams are used to convey a visual impression of the data and they must make the reading of the report a little easier. In the deluge of information to which we are subjected every day, concision is a virtue that is all too often forgotten, and concision associated with clarity is a goal that every one of us must strive to reach.

REVIEW QUESTIONS

The following questions review the most important concepts in every chapter. They constitute the basic structure of the knowledge acquired in this book. Therefore, it is not enough to know most of what is mentioned below: students must know it *all*. You should also refer to the checklist of abilities listed in the Foreword and at the beginning of every chapter.

Chapter 1

1. Explain what is a data file.
2. Explain the meaning of: case, variable, variable label, value, value label.
3. What do the terms *population, sample* and *unit* refer to? How about *element* and *case*?
4. What are the two main branches of statistics? Describe each briefly.
5. Give an example of a qualitative variable and one of a quantitative variable.
6. Explain the term level of measurement and list the three levels of measurement, describing the properties of the three types of measurement scales.
7. Both nominal and ordinal scales consist of categories. What are the two properties that are common to these two types of scales? What is the difference between them?
8. Describe the relationship between the type of variables and the measurement scales. (Class notes, in particular the diagram representing this relationship.)
9. Consider the variable Annual Income. Show three different scales for recording the data for that variable, one of which should be a numerical scale, one ordinal with many levels, and one ordinal with three levels.
10. Give an example (*other* than the one given in the book) of a how a complex concept could be broken down into dimensions and indicators.

Chapter 2

11. List the main steps of the research process.
12. Explain each of these steps briefly.
13. Explain these terms: research design; sampling design; experimental design.

14. Explain the three main research methods that produce quantitative data.
15. Explain the basics of experimental research (that is, the basic experimental research design).
16. There are four ethical principles that should guide any research with human subjects. List them and explain each briefly.

Chapter 3

17. What does the word univariate mean? What are the other, similar words you have seen in this book?
18. Explain the three types of univariate descriptive measures. Do measures of association fall into that category? What category do they fall into, then?
19. When you have to describe the data collected about a variable, the method used to describe it depends on one important characteristic of the variable: what is it?
20. What is the difference between a percentage and a proportion? What are the measurement scales for which percentages or proportions are appropriate descriptive measures?
21. What is the SPSS command that produces percentages of the various categories of the variable?
22. What is the difference between *percent* and *valid percent* in SPSS?
23. What is the difference between a majority and a plurality?
24. In the last election that was held in your country, did the winning party win a majority or a plurality of the popular vote? Did it win a majority or a plurality of the seats in parliament?
25. How do you find the median when there is an odd number of entries? How do you do it if there is an even number?
26. What does the expression: 5% *trimmed mean* mean?
27. What are the three quartiles? What is the interquartile range?
28. Write the equation for the standard deviation for a population and that for the standard deviation for a sample, and write the symbol used for each.
29. Suppose you have 140 men and 60 women in a group. What is the percentage of men in this group? What is the proportion of men in the group? What is the ratio of men to women?
30. When a distribution is symmetric, what can you say about the mean and the median? What if it is skewed to the right (that is, positively skewed)? How about negatively skewed?
31. What is the effect of extreme values on a distribution? How does the existence of such values affect the mean? and the median?
32. When the distribution is extremely skewed, which measure best represents the bulk of the data?

Chapter 4

33. What are the descriptive measures that are appropriate to use for a variable measured at the nominal level? What are the appropriate charts?
34. What are the descriptive measures that are appropriate to use for a variable measured at the numerical scale level? What are the appropriate charts?
35. What are the descriptive measures that are appropriate to use for a quantitative variable measured at the ordinal level? What are the appropriate charts?
36. If a quantitative variable measured at the numerical scale has been recoded into three categories, which measures are appropriate to describe the distribution?
37. What are the main components that should go into a descriptive report aiming at summarizing the data of a data file?

Chapter 5

38. A normal distribution is described by two numbers. What are they? What is the symbol used to describe a normal distribution?
39. What are the three basic characteristics that define a normal distribution?
40. What is a z-score?
41. Write the formula that transforms an x-score into a z-score, and the formula that does the opposite.
42. In $N(0,1)$ 95% of all data falls between ...... and ....
43. In $N(\mu,\sigma)$, 95% of all data falls between ...... and ....
44. In $N(\mu,\sigma)$, 90% of all data falls between ...... and ....
45. In $N(\mu,\sigma)$, 99% of all data falls between ...... and ....
46. In $N(\mu,\sigma)$, 5% of all data is larger than ....
47. In $N(\mu,\sigma)$, 1% of all data is larger than ....
48. In $N(\mu,\sigma)$, 2.5% of the data falls on each side, outside the following limits: ...... and ....
49. Draw a rough sketch to illustrate each of questions 42 to 48.

Chapter 6

50. Explain the terms *statistic* and *parameter*.
51. What is a sampling design?
52. What is a sampling frame?
53. Explain the difference between two broad categories of samples: probability samples and non-probability samples.
54. List the various types of samples.
55. Explain briefly each of these types.
56. Explain the procedure for selecting manually a simple random sample.

Chapter 8

57. Explain the main features of the notion of statistical association.
58. Explain the terms: dependent variable and independent variable.
59. Is the notion of independent variable intrinsic to a given variable? Explain.
60. Indicate the various procedures to be used to test statistical association, depending on the level of measurement of the dependent and independent variables.
61. Explain the notion of strength of a correlation.
62. The coefficient of correlation tells you about the strength of the statistical association if both variables are .... and the relationship is ....
63. Give an example of a scatter diagram where there is a strong statistical association, but where the association is not linear (in this case the coefficient r is misleading).
64. How do you analyze statistical association when the two variables are categorical?
65. Explain the difference between statistical association and relationship between variables.

Chapter 9

66. Give an example of a confidence statement involving a proportion.
67. Give an example of a confidence statement involving a mean.
68. List all the elements that must be part of a confidence statement and explain them.
69. Explain the difference between the margin of error and the probability of error.
70. For a given sample size, if we want to increase the confidence level, the margin of error then becomes .... (larger or smaller?)
71. Draw a diagram that illustrates the relationship between a point estimate and an interval estimate.
72. For a given confidence level, what happens to the margin of error if the sample size is multiplied by 4?

Chapter 10

73. Explain the difference between estimation and hypothesis testing (refer to the diagram on inferential statistics).
74. Write the three forms of formulation of the Null and Alternative Hypotheses.
75. Draw a diagram showing the acceptance region and the cut-off points.
76. Write the formula that allows us to compute the acceptance region for a two-tailed test at the 95% confidence level.
77. Write the formula that allows us to compute the acceptance region for a two-tailed test at the 99% confidence level.
78. Explain what are Type I and Type II errors.

LAB 1: GETTING STARTED WITH SPSS

The purpose of this lab is to learn how the windows of SPSS are organized, how to read the information displayed, and how cases and variables are organized.

1. Logging on

Turn the computer on (the CPU and the monitor). In most colleges and universities, the administrators of the computer network will provide registered users with a specific procedure to log on. You will eventually get a startup screen, with the **Start** button left of the screen.

2. Starting SPSS

The following labs have been written for SPSS 11.0. The dialog boxes for SPSS 10.1, for the SPSS Graduate Pack, and for SPSS 9.0 are almost identical. As for SPSS 8.0, the windows and dialog boxes look slightly different, but all the analyses shown in this manual can be performed with any of these versions. The main difference between version 8.0, and the subsequent versions is that there is no Variable View window in version 8.0. The statistical procedures used in this manual are almost identical in the various versions.

An SPSS session can be opened like any other Windows-based program. For those who are not familiar with this procedure, click and hold the **Start** button. A menu unfolds. Go up to the word **Programs** while you still hold down the left button on the mouse. The line becomes highlighted, and another menu opens. Move the mouse pointer until the words **SPSS 11.0 for Windows** become highlighted, then release the mouse button. The SPSS program will open, and you see the dialog box shown in Figure 1.1.

You will notice that you have several choices, including: creating a new SPSS document, or accessing more files. In the upper box, you see the words **More Files...** and then the list of SPSS files that have been opened recently. If the file you want is not in this list, select **More Files...** by clicking on it, and then click **OK**. You get the dialog box shown in Figure 1.2.

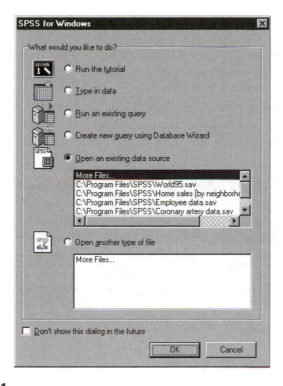

Lab 1 Figure 1.1

Lab 1 Figure 1.2

The dialog box gives you a list of the available files. We are going to work on the **GSS93 subset** file. Click on it once, you will see the name of the file appearing in the **File name** box. Click **OK**. You should now get the **GSS93 subset** data window open, as shown in Figure 1.3. Every time you open an SPSS data file, you get what is called an **SPSS Data Editor**. It can appear in one of two views: a **Data View** and a **Variable View**. At the bottom left of the SPSS window, you see two little tabs that allow you to display one or the other of the two views. Figure 1.3 displays the Variable View. In this view, every row contains the information that

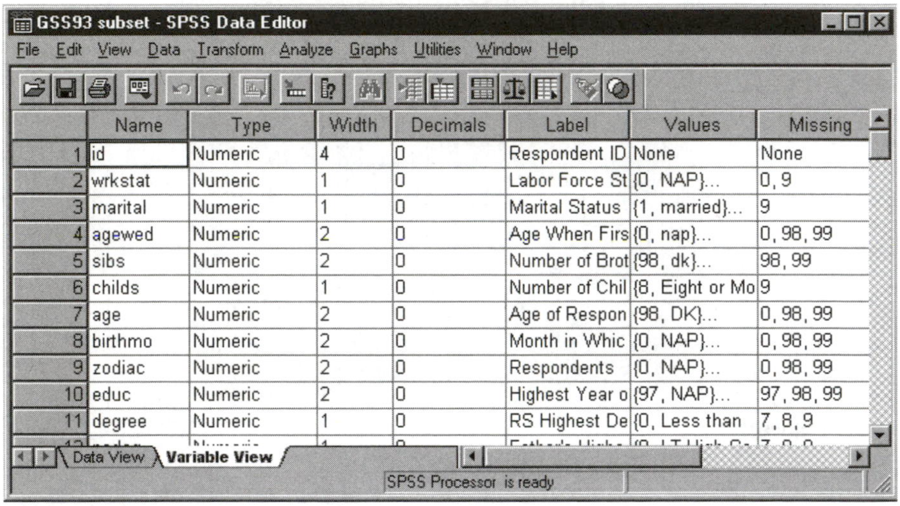

Lab 1 Figure 1.3

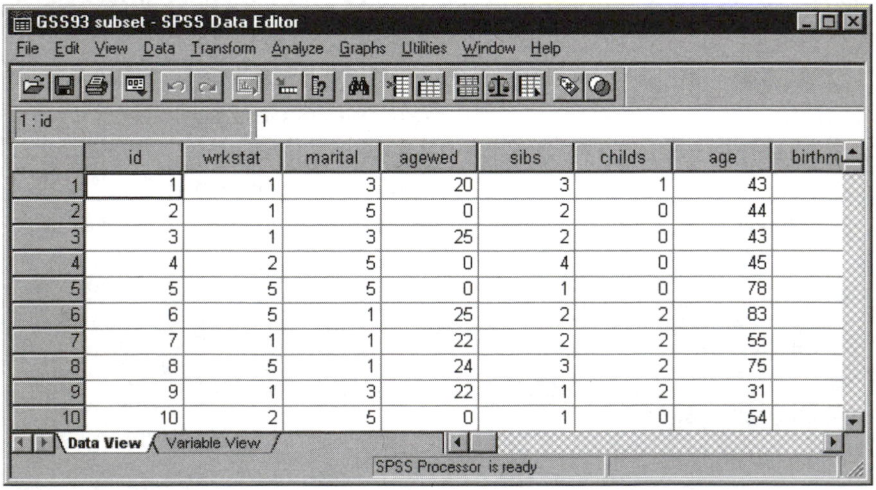

Lab 1 Figure 1.4

pertains to one variable. We will explain this information after we look at the other view, the Data view.

When you click on the Data View tab of the **SPSS Data Editor**, you see the data itself, and you can modify it directly in this window (Figure 1.4). Recall now all the definitions learned in the Chapter 1 of this book, as we will refer to them in the exploration of SPSS that now follows. Perform the operations indicated below and see their effect on the screen:

1. Get SPSS to display the **Value Labels** instead of the codes by selecting Value Labels under View.
2. Read the full name of any variable by pointing the mouse on the short name, at the top of any column.
3. Enlarge any column by positioning the mouse right on the edge separating the variable names, and drag slightly to the right without lifting your finger.
4. Select **Variables** under the **Utilities** menu. You will get the dialog box shown in Figure 1.5. By scrolling down the list of variables with the mouse, or simply moving by using the arrows on your keyboard, you will be able to see the detailed description of each variable, one at a time.

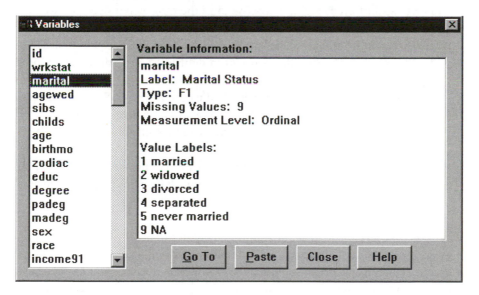

Lab 1 Figure 1.5

You will see:

*   the **Variable Label** (which is the long name of the variable);
*   the type and the missing values, which will be explained further down;

- the **Value Labels**, and their corresponding codes; this is what tells you that

  1   stands for married
  2   stands for widowed, etc.

5. The **Type**. This is the format of the numerical values or labels you enter.

   F1   means that the format used for that variable is one space and no decimals.

   F4.2 means that 4 spaces are reserved to write the entries of that variable, of which 2 are decimals. The number 3.62 has the format F4.2, because 4 spaces are needed to write it (the dot uses a space).

   F6.1 is a format that uses 6 spaces, including one decimal and the dot; Ex. 4527.3.

   Scroll down the variable list, examine the various types that are written, then compare them with what you see in the Data View window.

6. **Missing values.** For each variable, some of the answers should not be taken into account in the statistics, such as when somebody refuses to answer, or when the question does not apply to that person. We give codes for the values 'Refuses to answer' and 'Does Not Apply', but we must indicate that these answers are not to be treated like the other answers. We label them Missing values. We will see later on how to do that.

7. **Measurement level.** This refers to what we have seen in Chapter 1. In this data file, the variable **Marital Status** has been defined to be Ordinal. The proper measurement level for this variable should have been Nominal. Sometimes, however, nominal variables are labeled Ordinal to allow SPSS to perform certain operations that can be done only on ordinal or on numerical scales.

8. **File Info.** If you go under the menu **Utilities** and select **File Info**, you will get all the information about all the variables in the file in one shot. A new window will be opened, call **Output1 SPSS Viewer**, shown in Figure 1.6. The **File Info** command produces what we called the *codebook*. It contains all the information needed to interpret correctly the data found in the file.

Whenever you give a command in SPSS, the output will be displayed in the SPSS Viewer. You can save the output of a whole SPSS session on your diskette and give it a name. You can also print the whole output (which may be very long if you have been working for a while) or only one part of it. To print just one part click on it: it becomes surrounded by a rectangle, which means it is selected. You can then print the selection by giving a print command. You will learn in Lab 2 how to copy elements from the output window and paste them into a word processing document.

The three little boxes that are on the top right are no doubt familiar to most readers. They have the following functions. By clicking on the first one, you *minimize* the output window, which means it appears as a small rectangle on the bottom of your screen. By clicking on that rectangle, you open the window again. The second box changes the size of the window, and the third one closes the window altogether, after asking you if you want to save its contents. Try each of these operations. Notice that

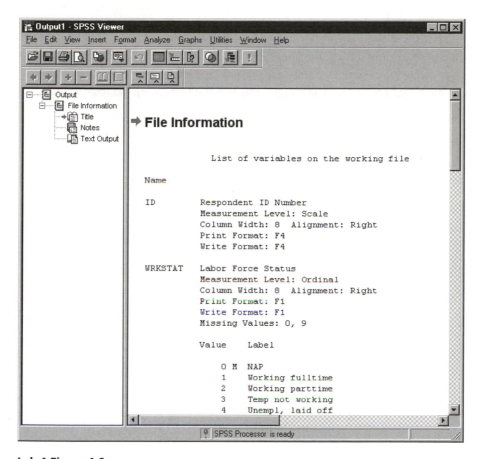

Lab 1 Figure 1.6

the output window is split in two vertically. The part on the left is called the *document map*. When you have produced a large output, the document map allows you to move quickly from one element to the other, simply by clicking on the little icon representing the table or graph that you want to see. The following labs will help you become familiar with these various functions. Finally, there is a third type of window in SPSS, called the Syntax window. An SPSS syntax is a set of commands which are written in a special code, called the syntax. The syntax allows us to perform more operations in SPSS than with the menus, and in a much more efficient way. The syntax needed to obtain a frequency table for the variable Marital Status is, for example:

```
FREQUENCIES
  VARIABLES = marital
  /ORDER = ANALYSIS.
```

The syntax must follow very precise rules that are explained in the SPSS online manuals. We will not be working with syntax systematically in this book, but you should know what it is and we will make occasional remarks on the syntax. You should know that the great advantage of syntax is that when you have a long set of commands you can modify any command manually (for instance we can add another variable next to the word marital), and you can repeat the same analysis with or without modifications just by selecting all commands and clicking the Run button.

LAB 2: WORKING WITH A WORD PROCESSOR

The purpose of this lab is to learn how to copy tables and graphs from SPSS into a word processor and how to do a minimum amount of editing. This lab is intended for those who have no familiarity with word processing software and should be skipped otherwise.

We will learn how to create a simple document with a word processing software, and how to incorporate in it selected parts of the SPSS output. We will be using MS Word, but the principles used here are the same for any word processing software such as WordPerfect which is also widely used. Try each one of the operations suggested below in order to familiarize yourself with them, as you will have to type your analysis of an SPSS file and insert in your document tables and graphs taken from the SPSS output. The actual commands may differ a little depending on how the local setup of the computer lab is organized. The explanations below will work for most situations.

1. **Opening Word and writing**
   Click with the left button of the mouse on the **Start** button on the bottom left of the screen. With the button pressed down, select **Programs** then **Microsoft Word**. Start writing any sentence.
2. **Writing your name in the header**
   The *header* is a paragraph that is printed automatically at the top of every page. You can write your name and the title of the document you are working on. You can also have the page numbers included, and Word will automatically number the pages correctly. To open the header, select **Header and Footer** under the **View** menu. The header will appear in a dotted line, and a tool bar will also appear. Write

   Lab 2 – your name.   p.

   To paginate your document, use the Tab key on your keyboard to place the cursor to the right of the header, the type: 'p.' and click on the little icon in the tool bar that has a number sign (#) in it. The correct page number will now automatically appear on every page in your document, next to the lab number and to your name. Click on **Close** in the tool bar.

3. **Saving a document**

   Insert your disk in the floppy disk drive. Select **Save** under the File menu. The program offers to save in some folder (usually the folder labeled **My Documents**). Click on the little triangle that appears on the right end of this dialog box. A list of folders will be displayed; click once on the **3½ Floppy (A:)** icon. Give a name to your document and click on **Save** (the program will spontaneously suggest the first line of your text as the name of your document). This will save your document on your floppy disk. The **Save** operation can also be performed by clicking on the floppy disk icon shown in the toolbar above the ruler.

4. **Opening a saved document**

   Open Word, then select **Open** from the **File** menu. When the window appears, click in the little triangle at the right end of the dialog box called Look in: at the top of the window in order to see and select your floppy disk on Drive A.

5. **The difference between Save and Save as...**

   When you save a document for the first time, there is no difference between **Save** and **Save as**. When a document has been already saved then modified, the **Save** command will replace the old document by the modified document, incorporating all the changes you have made. The old document will be lost. The **Save as...** command will give you the possibility to change the document name and save it under a new name, while keeping the original intact. Any small change (such as adding 2 to the name of the document) is considered a new name.

6. **Inserting new text in the middle of existing text**

   Just put the cursor where you want to insert your text, and type. You can put the cursor there either by clicking on the mouse, or by using the arrows on your keyboard. If you find that the new letters you type erase the old ones and replace them, press the key labeled **Insert** on your keyboard: this will change the insert mode from *insert by replacing* to *insert new text*.

7. **Modifying your typed text by the commands Cut, Copy or Paste**

   When you select a piece of text by dragging the mouse across it, you can perform certain operations like cutting it or copying it. Then you can insert this text anywhere in your document by positioning the cursor in the new location, then giving the **Paste** command. The three commands are under the **Edit** menu.

8. **Inserting tables of graphs from SPSS into a Word document**

   You can have both SPSS and Word opened at the same time. You move from one to the other by clicking on the appropriate rectangular bar at the bottom of your screen. When you are in the SPSS Output window, you can select any table or chart by clicking on it. You can then click on **Copy** under the **Edit** menu (you could also press Ctrl C simultaneously). Now go to your Word document, place the cursor where you want the table pasted, then click **Paste** under the **Edit** menu (or press Ctrl V).

9. **Two modes of copying a table from SPSS**

   There are two ways of copying a table from SPSS into Word. When you have selected the table, you can choose either **Copy** or **Copy Object** under the **Edit** menu. The difference is the following:

Copy will copy the *contents* of the table. When you paste it in Word, you can edit it and reformat it (that is, you can change the way it looks). For instance you can select all the rows of the table, then click **Table Autoformat** under the **Table** menu, and select an automatic format. This operation of automatic formatting of a table is possible only if you select all the rows of the table, and nothing more.

Copy Object will copy the table as is, with its borders, and when you paste it in Word you cannot edit it: you cannot delete any row, and you cannot change the content of the table.

10. **Printing part of a document**

When you give a **Print** command under the **File** menu, you can indicate the range of pages to be printed, and you can even choose to print only the selected text (if a text is selected in Word). This is useful to avoid printing pages uselessly.

EXERCISE

Open both Word and SPSS. Select any data file in SPSS (for instance the **GSS93 subset** file), and produce its codebook by giving the **File Info** command under **Utilities**. In the Output window, click on the **File Info** to select it, then copy it and paste it in Word. Add at the top of the page the name of the file it comes from. Write Lab 2 and your name in the Header, and insert the automatic page numbers. Then print ONLY PAGE ONE of your document, and show it to the instructor for comments. Keep the printed page with your labs.

LAB 3: EXPLORING DESCRIPTIVE STATISTICS

The purpose of this lab session is to get you acquainted with the most common procedures for describing data with SPSS.

We will be working with the data file **GSS93 subset** that is provided with the SPSS program. In order to become familiar with the procedures available in SPSS, we will answer questions such as:

- What is the average age of people when they first marry?
- What is the average age of men in this sample?
- What is the average age for women?
- Can you give a visual description of how the variable age is distributed?
- What is the proportion of people who favor capital punishment?

There are four commands that produce descriptive statistics, all under the **Analyze →
Descriptive statistics** submenu, as shown in Figure 3.1. These procedures are:

>the **Frequencies…** command,
>the **Descriptives…** command,
>the **Explore…** command,
>the **Crosstabs** command, and
>the **Ratio** command.

In each case you have certain options that allow you to obtain specific statistics or to produce visual summaries (graphs and charts). This first lab on descriptive statistics will be an exploratory one: you will learn how to give some simple and straightforward commands. In subsequent labs, we will learn how to select the appropriate measures.

Warning: SPSS is a very powerful program, which provides a very wide range of statistical procedures. We will not use them all. You will have to focus on the options you know, and ignore the others. Do not use outputs that you do not know how to interpret, and delete the ones you produce accidentally.

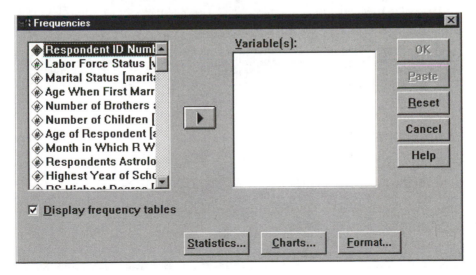

Lab 3 Figure 3.1

The Frequencies...Command

1. Select **Frequencies...** by clicking on it.

    You get the dialog box shown in Figure 3.2. The procedures explained
    below are more adequate when the variable is *qualitative*. However, they

Lab 3 Figure 3.2

are also useful when the variable is *quantitative but grouped into a small number of categories*, as with the variable **Age Categories [agecat4]**. All the variables are listed in the box on the left. If you want to compute any statistic, you must first select the variables with which you want to work, one at a time, and place them in the box on the right by clicking on the arrow.

2. Select the variables **Marital Status** and **Age Categories** (not **Age of Respondent**; the former variable is placed towards the end of the list of variables) and place them in the box on the right.

    You must then indicate what statistics you want SPSS to compute. You can have the Frequencies tables displayed (this is the default option), and that is useful when the level of measurement is nominal or ordinal, with a small number of categories. If your variable is continuous, remove the check mark next to **Display frequency tables**. For the two variables in this exercise, leave the check mark as it is.

3. Click now on the **Statistics...** button.

    You get the dialog box shown in Figure 3.3. You have four boxes within that dialog box, each displaying one type of descriptive statistics. Refer to the chapter on descriptive statistics to review the definitions of these measures. Only the **Mode** is useful among the measures of central tendency, since our two variables are measured at the nominal level. None of the other measures are useful at this point. Click the box next to the word **Mode**.

4. Click **Continue**; you will get back to the preceding window.

Lab 3 Figure 3.3

5. Click now the **Charts...** button in the **Frequencies...** window. You get the dialog box shown in Figure 3.4.

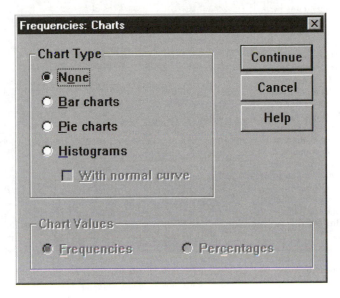

Lab 3 Figure 3.4

You can choose among three types of charts: **Bar charts, Pie charts**, and **Histograms**. Once you have chosen a chart type, you must indicate whether you want the frequencies or the percentages to be displayed.

6. Click **Bar charts,** select **Percentages**, and click **Continue**.
7. In the **Frequencies...** window you come back to, click **OK**.

Exercise 1

Examine the display you get, and write one sentence for each of the two variables that summarizes what you see on the charts and in the tables.

Exercise 2

Re-do the same exercise and select **Pie Charts** instead. Re-do it with **Histograms**. Write a paragraph to explain the advantages of each type of display for our two variables.

Exercise 3

Produce a clustered Bar Chart by performing the following steps:

1. Select **Bar...** under **Graphs.**

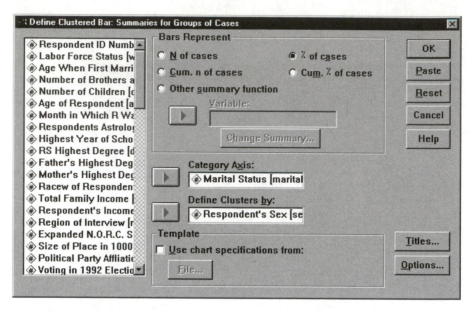

Lab 3 Figure 3.5

2. In the dialog box you get, click the square button for **Clustered** and the little round button to select **Summaries for groups of cases**.
3. Click **Define**. You get the dialog box indicated in Figure 3.5.
4. Place the variable **Marital status** in the **Category Axis**: rectangle and place the variable **Respondent's sex in the Define Clusters by:** as shown in Figure 3.5.
5. Make sure that in the box labeled **Bars Represent** in the top part of the dialogue box, you select **% of cases** rather than **N of cases**.
6. Now click **OK**.

You get a clustered bar graph as explained in Chapter 3. Which are the Marital categories where there is a higher percentage of men than women? Which are the categories where the percentage of women is higher? Can you think of a reason why this is the case?

Interpreting the SPSS Output

The preceding procedure gives you one frequency table and one chart for each variable selected, but it is preceded by an additional table which includes all the variables selected. That first table gives you, for each variable, the number of **valid** answers and the number of **missing** answers.

VALID ANSWERS

In the preceding examples, you can read that the number of valid answers for the variable **Attended Sports Events Last Yr** is 1489. That means that we do not have the answers for 11 individuals: either they did not know or did not remember, or they refused to answer, or the question did not apply to them. The subsequent table will give us such details. We also see in the same table that we do have all the 1500 answers for the variable **Respondent's Age**.

FREQUENCY TABLES

Examine the frequency table carefully and write down the kind of information found in each of its columns in the space provided below.

Column 1:   _____

Column 2:   _____

Column 3:   _____

Column 4:   _____

Column 5:   This column contains the cumulative percentages, which are useful for quantitative variables only.

Now examine the rows. The first few rows correspond to the *categories* of your variable (one row for each category: Yes/No; 18–29/30–39/40–49/50+). They are followed by the total number of *valid* answers. You must pay attention to this row, specially the *percentage* of valid answers, If a small percentage of answers are missing, it does not affect the results very much. But if the percentage of missing answers is high, you must ask yourself whether the conclusions you draw from the table are well founded. The next rows give the details of the missing answers:

DK stands Don't Know;

NA stands for No Answer;

NAP stands for Not Applicable;

the number and percentage of people who are in each of these categories is listed, unless there are none.

THE BAR CHART

Look at the bar chart for **agecat4**. You notice that the highest bar is for the category 50+. Does it mean that there are more old people than young people in this sample? This is not necessarily so. The point is that the range of the category 50+ covers close to 40 years, whereas the range of the preceding category (40–49 years) covers only 10 years. This is why there are more people in the category 50+. If the ages were grouped into equal intervals of 10 years each, the picture would have been different. Review the discussion of class intervals in the chapter on descriptive statistics.

The Descriptives... Command

The **Descriptives** command is adequate for *quantitative variables only*.

1. Select the **Descriptives...** command (**Analyze** → **Descriptive Statistics** → **Descriptives...**).
2. Determine the variables you want to analyze and place them in the **Variables** box on the right. Do that for the variables **Age of Respondent** and **Age when First Married**.
3. Click on the **Options** button to indicate which statistics you would like to see displayed. You get the dialog box shown in Figure 3.6. Notice that no charts are offered as an option in this command: just the basic descriptive statistics. Keep this default option.
4. Click **Continue**. You get back to the first dialog box with your two variables written in the Variables box.
5. Click **OK**.

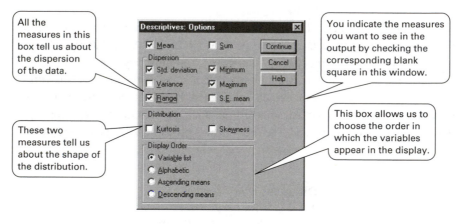

All the measures in this box tell us about the dispersion of the data.

You indicate the measures you want to see in the output by checking the corresponding blank square in this window.

This box allows us to choose the order in which the variables appear in the display.

These two measures tell us about the shape of the distribution.

Lab 3 Figure 3.6

The output of the **Descriptives...** command is a single table that includes all the statistical measures you have checked in the **Options** window. Examine the first column in the table, whose heading is **N**. You read in that column the number of valid answers you get for each variable. For the variable **Age of Respondent**, for instance, the valid N is 1495. But you also get in that table something called **Valid N (list-wise)**. This is the number of individuals for whom you have answers *for all the variables described in that table*. It is important to look at this number when you want to compare the way the same person answers different questions.

Lab 3 Figure 3.7

The Explore...command

The **Explore...** command (**Analyze → Descriptive Statistics → Explore ...**) gives us information not only on descriptive measures, but also on some other measures that belong to inferential statistics and that will be studied later in the course. We will ignore such measures for the time being. This command is a little more complex than the previous one, as it offers many options. *It applies only to quantitative variables.*

1. Click on the **Explore...** command (**Analyze → Descriptive Statistics → Explore...**). You get the dialog box shown in Figure 3.7.
2. Select the variable **Age When First Married** and place it in the **Dependent List** box by clicking on the button with a little arrow. This is where you place the variables that you want to analyze, and such variables *must be quantitative.* Leave the other boxes blank for now.
3. Click **OK**.

The output of the **Explore...** command for that variable is shown in Figure 3.8. Ignore the rows and the column that are shaded in dark gray, as they will be explained in the chapter on inference.

Interpretation of the Output of the Explore... Command

The first table contains information about the number of valid and missing answers. All the statistics in the second table, except the ones that have been shaded, have

Case Processing Summary

	Cases					
	Valid		Missing		Total	
	N	Percent	N	Percent	N	Percent
Age When First Married	1202	80.1%	298	19.9%	1500	100.0%

Descriptives

			Statistic	Std. Error
Age When First Married	Mean		22.79	.145
	95% Confidence Interval for Mean	Lower Bound	22.51	
		Upper Bound	23.08	
	5% Trimmed Mean		22.40	
	Median		22.00	
	Variance		25.331	
	Std. Deviation		5.033	
	Minimum		13	
	Maximum		58	
	Range		45	
	Interquartile Range		6.00	
	Skewness		1.658	.071
	Kurtosis		5.382	.141

Lab 3 Figure 3.8

been explained in the chapter on descriptive statistics: review these notions to be able to interpret this output correctly. One of the important uses of the **Explore...** command is that it allows you to compute the descriptive statistics separately for subgroups of the population, such as: men vs. women; various age categories, etc. The exercise below will illustrate these computations.

Exercise 4: the Explore... Command with Subdivisions by Groups

If we want to obtain the average age at first marriage for men and women separately, we must place the variable **Age When First Married** in the **Dependent List** box as before, and place the variable **Respondent's Sex** in the **Factor List** box. The output will give you the statistics for men and women separately. To answer the questions below, you must place some quantitative variable in the Dependent List box, and a qualitative variable in the Factor List box. You must decide which are the appropriate variables.

> **IMPORTANT**: The variables placed in the **Dependent List** box must be quantitative. The variables placed in the **Factor List** box must be either nominal or grouped into a small number of categories.

Answer the following questions by producing the **Explore...** command output as explained above:

The average age at which men marry for the first time: _____
The average age at which women marry for the first time: _____

Give the **Explore...** command using two new variables: **Age of respondent** in the Dependent List box and **Marital Status** in the Factor List box and find the following statistics:

The average age of the respondents who are married: _____
The average age of the respondents who are widowed:_____

Repeat the procedure using **Age of respondent** and **Married?** instead of **Marital Status**. Write a full sentence that expresses the results that you have obtained.

LAB 4: RECODING; THE HELP MENU

The purpose of this lab is to learn how to recode variables into new variables (usually to create a smaller number of categories), and how to use the Help menu.

We will be using the **GSS93 subset** data file for all the exercises in this lab session. Open the **GSS93 subset** data file.

Recall that the purpose of descriptive measures is to convey, in a nutshell, essential information about the data that relates to a variable. When you want to examine a variable, you must first recall what its name stands for, and what the value labels are. You can find this information in the codebook, which is produced by the command **File Info** under the **Utilities** menu. A simpler way to see it is the command **variables...** under **Utilities**. It displays the information for one variable at a time, and it is easier to consult (you do not have to scroll down for ever...). Moreover, the command **variables...** does not place the info in the Output window, which is better when you to not want to print this information. Once you have recalled what the variable stands for and what its value labels are, etc., you may want to examine the data that relates to that variable: its summary statistics, which include measures of central tendency, measures of dispersion, measures of position, and measures that concern the general shape of the distribution of that variable.

Practice for the Command Frequencies

Exercise 1

Review the commands needed to produce frequencies, seen in Lab 3. Produce the frequency tables for the variables **Marital Status, Labor Force Status, Should Marijuana Be Made Legal** and **Number of Children**, and answer the following questions. You may have to add the percentages given in the SPSS output to answer some of the questions.

Q1. How many missing answers are there for the variable **Number of Children?**

_____

Q2. How many valid answers are there for the variable **Marital Status**?

_____

Q3. How many valid answers are there for the variable **Labor Force Status**?

_____

Q4. How many people are divorced? _____

Q5. What is the total percentage of people in this sample who said marijuana use should be made legal? _____

Q6. What is the valid percentage of people in this sample who said marijuana use should be made legal? _____

Q7. What is the percentage of people working full time? _____

Q8. What is percentage of people working either part time or full time? _____

Q9. How many people have two children? _____

Q10. What is the percentage of people who have more than three children? _____

Frequencies of a Quantitative, Ungrouped Variable

Produce the frequency of the variable **Age of Respondent**. You will get a very long table, with one line for every age occurring in the sample. The frequencies themselves are not very useful to describe this distribution. However, the cumulative frequencies are useful: you can read off directly how many people are less than a certain age. One of the options in the Frequencies command allows you to have a decreasing frequency distribution, which would let you see directly the percentage of people that are older than any given age.

Q11. What is the percentage of people who are 45 years old or younger? _____

Q12. What is the percentage of people who are between 40 and 44 years old, including both values? (you need to make a subtraction) _____

Use of the Recode Command

Notice that the ages are not grouped in categories for the frequency table of the variable **Age of Respondent**. Your analysis may require specific answers about, say, the 18–30 age group, versus the retired persons. To do such an analysis, you may want to group the ages into categories. This SPSS file does contain a variable called **Age Categories** (agecat4) which presents the data on age grouped into four categories. We will now learn how to perform similar operations. The procedure used is called **recoding**. This is how we can recode the age values to produce a frequency table of grouped values.

You must first decide how you want to recode your variable. We will recode it into six categories as follows.

Category	Code
18 to 30	1
31 to 40	2
41 to 50	3
51 to 60	4
61 to 70	5
71 and over	6

1. From the Data view of the data editor, select **Transform** → **Recode** → **Into Different Variables...** You get the dialog box shown in Figure 4.1.

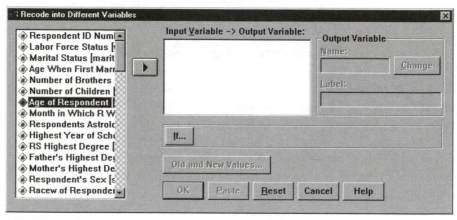

Lab 4 Figure 4.1

2. Select the variable **Age of Respondent** from the list of variables on the left, and place it into the box **Input Variable→Output Variable**.
3. Type a new name in the **name** box (this is the short name that will appear in the data window, on the top of the new column). We will call it **age6**.
4. Give it also a **Label** (this is the long name that will appear in the codebook). Type **Age grouped into 6 cat** in the Label box.
5. Click **Change**. The new name will appear next to the old name in the central box.
6. Click the button **Old and New Values**. You get the dialog box shown in Figure 4.2. This is where you are going to define the six categories you want for your new variable.

Take a minute to examine the various choices offered for recoding in the dialog box. For each of the new categories, we will determine its range in the **Old Value** box, then assign a new value to it in the **New Value** box, then click **Add** when we are sure we have defined it correctly. Here are the details.

7. Select the first occurrence of the word **Range** by clicking in the button that precedes it.
8   Type the first range of values: **18** through **30**.
9. Type **1** in the **Value** box.

Recode into Different Variables: Old and New Values

Old Value	New Value
● Value: [ ]	● Value: [ ] ○ System-missing
○ System-missing	○ Copy old value(s)
○ System- or user-missing	Old --> New:
○ Range:	[Add]
[ ] through [ ]	[Change]
○ Range:	[Remove]
Lowest through [ ]	
○ Range:	☐ Output variables are strings Width: [8]
[ ] through highest	☐ Convert numeric strings to numbers ('5'->5
○ All other values	[Continue] [Cancel] [Help]

Lab 4 Figure 4.2

Recode into Different Variables: Old and New Values

Old Value	New Value
○ Value: [ ]	● Value: [ ] ○ System-missing
○ System-missing	○ Copy old value(s)
○ System- or user-missing	Old --> New:
○ Range:	[Add] 18 thru 30 --> 1
[ ] through [ ]	31 thru 40 --> 2
	[Change] 41 thru 50 --> 3
○ Range:	51 thru 60 --> 4
Lowest through [ ]	[Remove] 61 thru 70 --> 5
● Range:	71 thru Highest --> 6
[ ] through highest	☐ Output variables are strings Width: [8]
	☐ Convert numeric strings to numbers ('5'->5
○ All other values	[Continue] [Cancel] [Help]

Lab 4 Figure 4.3

10. Click **Add**. You should see now the phrase: **18 thru 30 → 1** appearing in the Old → New box.
11. Repeat this operation for each of the new categories. For category 6, you will have to select the button **Range: _____ through highest** and then type 6 then click Add. You should now see the dialog box shown in Figure 4.3.
12. Click **Continue**. You get back to the previous dialog box (shown in Figure 4.1, but now with all the blanks filled out).
13. Click **OK**. You have now created a new variable **age6**, labeled **Age grouped into 6 cat**, which is placed at the end of the list of variables.

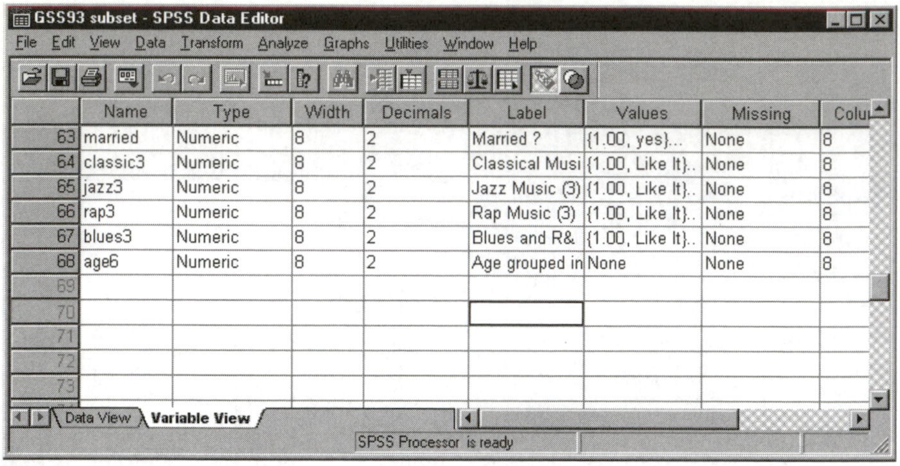

Lab 4 Figure 4.4

Lab 4 Figure 4.5

But that new variables does not have value labels yet. This means that when you produce a frequency table, the age categories will appear as 1, 2, 3, etc., but it will not be indicated what 1, 2, 3, etc. stand for. To indicate what these values stand for, click on the Variable View of the data editor, and scroll down to the end of the list of variables. Your new variable should now be listed, as shown in Figure 4.4. In order to put labels to the values of that variable, do the following.

14. Click once on the right side of the cell corresponding to the Values of age6, that is on the cell sitting on the same line as **agecat6**, and in the column **Values**. The word None appears in the cell before you click on it. But when you click on its right side, the dialog box shown in Figure 4.5 now appears.

15. Type **1** in the Value box, type **18 to 30** in the Label box, and click **Add**.

16. Repeat this operation for each of the six age categories.

17. Click **OK** when you are finished. You have now defined the labels of six categories.
18. Click in the Decimals column for age6, and change the 2 to a 0 as we do not need decimals.
19. Save your data file on your disk to be able to use the recoded variable in the future.

Exercise 2

Produce a frequency table for the variable **Age grouped into 6 cat**.

Q13.   What is the number of people who are in the 31–40 years old category? _____

Q14.   What is their percentage? _____

The Help menu

Before going further, you may want to gain some familiarity with the **Help** menu of SPSS. Under **Help**, three submenus are particularly useful: **Topics**, **Tutorial**, and **Statistics Coach**.

Lab 4 Figure 4.6

The **Topics** menu produces a list of all the topics dealt with by SPSS. By clicking on a topic, you can read some explanations, including definitions of the technical statistical terms, and practical indications on how to perform certain operations. For example, Figure 4.6 illustrates the list of topics that appear Under Statistical Analysis and then under Descriptive Statistics.

If you click on Frequencies, you get the window shown in Figure 4.7. Notice that you have three buttons on the top of that window, labeled respectively **How to, Syntax**, and **See Also**:

- The **How to** button tells you where to click in the menus to execute the procedure. By clicking on it you get the window shown in Figure 4.8.

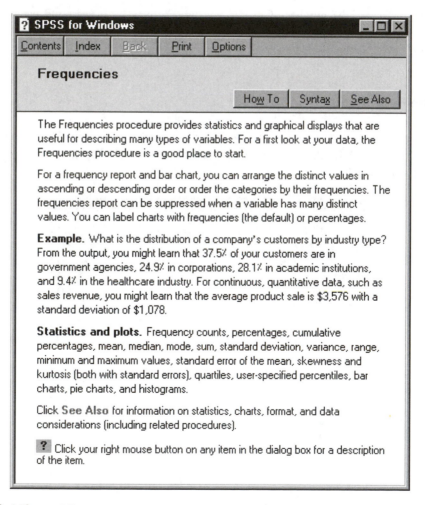

Lab 4 Figure 4.7

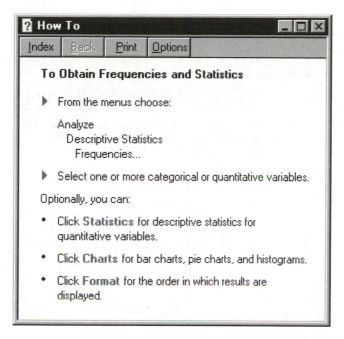

Lab 4 Figure 4.8

- The Syntax button tells you the rules governing the writing of the syntax if you choose to give the commands through the syntax window rather than through the menus.
- And the **See Also** button takes you to related topics.

Take a few minutes to examine some of the topics that have been seen so far.

The **Tutorial** menu is a systematic guided tour of SPSS. It illustrates in a visual way how to perform various operations in SPSS.

Take a few minutes to follow the Tutorial on some of the topics we have covered so far, such as **Select a procedure from the Menus**, which is listed under **Creating Tables and Calculating Statistics**.

The **Statistics Coach** asks you to determine a question, then a sub-question, and yet another one, and tells you at the end what is the statistical procedure you should perform to answer your question. It is more useful for those who know SPSS already but are not too sure of the analysis they should do for their specific question. It will become more useful towards the end of this course.

The Charts Gallery in SPSS

For **graphs**, there is a specific help under the **Graphs** menu. It is called **Gallery**. It shows all the possible types of graphs, and offers help to explain each type. When you select **Gallery** under **Graphs**, you get the window shown in Figure 4.9.

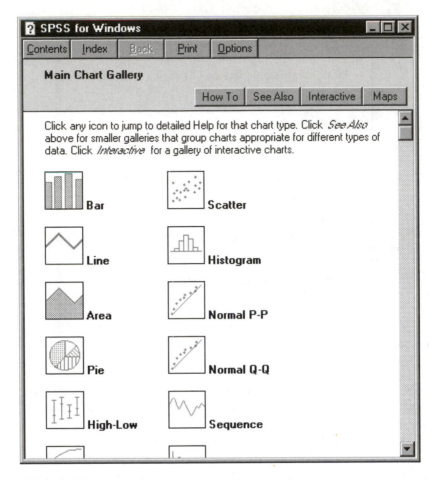

Lab 4 Figure 4.9

If you scroll down the window, you will see more chart types. By clicking on the icon of any chart type, you will get detailed indications on when to use it and on how to produce it.

Take a few minutes to explore the explanations given on the various chart types.

The following lab (Lab 5) will help you become familiar with the basic charts used in descriptive statistics.

LAB 5: CHARTS IN SPSS

The purpose of this lab is to show how to produce the main types of chart in SPSS.

Bar Charts

To produce a bar chart in SPSS follow the following procedures.

1. Select **Bar ...** under the **Graphs** menu.
2. In the dialog box you get, select **Simple**, and **Summaries for groups of cases**, then click **Define**.
3. Select the variable **Martial Status** and place it in the box labeled **Category Axis:**. In the box that says **Bars Represent,** click on the button that says **N of cases**. Click **OK**. Examine what you get in the output window. Repeat the procedure but select **% of cases** instead of **N of cases**.

Q1. What is the difference between the two bar graphs?

_____

_____

   Use the **clustered** bar charts to illustrate the percentages of men vs. women in the various educational categories.

1. Select **Bar...** under the **Graphs** menu.
2. In the dialog box you get, select **Clustered** and **Summaries for groups of cases**, then click **Define**.
3. Place the variable **RS Highest Degree** in the **Category Axis** box, and the variable **Respondent's sex** in the **Define clusters by** box. Click **OK**.

Q2. Make a short comment about the distribution thus illustrated (for instance, indicate which category contains a higher percentage of men or of women).

_____

_____

_____

_____

Pie Charts

1. Select **Pie chart** under the **Graphs** menu.
2. Select **Summaries for groups of cases**, and click **Define**.
3. In the dialog box that appears, select the variable **Marital Status** and place it in the box labeled **Define Slices by:**. In the box labeled **Slices Represents** select either **N of cases** or **% of cases**. Click **OK**.

You get a pie chart in the output window. Go back to the sequence of commands given above and find the option that allows you to include or exclude the missing cases.

Q3. What is the advantage or disadvantage that you see in including or excluding missing values?

Note: If you use Pie Chart on a variable such as Age of Respondent, you will get something totally useless, as there are far too many categories. Pie charts are used when the number of categories is small. The slices represent the relative importance of each category.

Q4. In your opinion, does the **Bar Chart** or the **Pie Chart** give a better idea of the distribution of the variable **Marital Status**? What advantages do you see in each type of graph?

_____

_____

Histograms

Histograms are more adequate than pie charts when the variable is quantitative: The bars represent a succession of values in a given order, whereas the pie chart is more adapted for qualitative variables where the categories are separate and are not ranked in a given order. To produce a histogram, do the following:

1. Select **Histogram** under the **Graphs** menu.
2. Place a quantitative variable in the appropriate space, such as Age of Respondent.
3. Click OK.

SPSS automatically divides the total range of a quantitative variable into a small number of classes of equal length. This is not done in the Bar Chart command. The difference between Bar Chart and Histogram is the following. A bar chart is more adequate when the variable is qualitative: the bars represent categories that are not ordered or ranked. A bar chart could also be adequate when a quantitative variable has been regrouped into a small number of categories. A histogram is better suited for quantitative variables: the bars represent successive values that are grouped into classes.

Q5. Draw histograms for the variables **Age of Respondent** and **Age grouped into 6 cat.** Compare the two. For the histogram of Age of Respondent, what are the midpoints of the various classes? Can you figure out the class boundaries (that is, the values that separate the various classes) directly from the chart? List all the classes that appear on the histogram you have just produced:

Number of classes used in the histogram for **Age of Respondent**: ____

Boundaries of the first four classes:

1. from ___to___   2. from ___to___
3. from ___to___   4. from ___to___

Numbers of classes used for the histogram for **Age grouped into 6 cat:** __
Class boundaries:
1. from ___to___
2. from ___to___   etc.

For which of these two variables is the histogram more appropriate? _____
Draw a bar chart for **Age grouped into 6 cat**. Which representation is better for this variable: the bar chart or the histogram?

Box Plots

Box plots are used to illustrate the distribution of quantitative variables. You can use them specially for comparisons: either to compare two comparable variables (for instance, Income in a given year, and Income the following year) or to compare how two subgroups are distributed for the same variable.

   Suppose we want to compare the age at first marriage for men and women separately and compare them. We will produce a box plot chart as follows.

1. Select **Box Plot** from the **Graphs** menu, then select **Simple** and **Summaries for groups of cases**, and click **Define.**
2. In the dialog box that you get, place the variable **Age When First Married** in the **Variable** box, and place the variable **Respondent's Sex** in the **Category Axis** box. Click **OK**.

You get two box plots in the same diagram that you can compare.

Q6. Describe the difference in distribution for the variable **Age When First Married** for men and women, and indicate the five-number summary for each group.

_____

_____

Clustered Box Plots

You may want to compare the age at first marriage of men and women across other categories such as the highest degree attained. Use the following procedure.

1. Select **Box Plot** from the **Graphs** menu, then select **Clustered** and **Summaries for groups of cases**, and click **Define.**
2. In the dialog box that you get, place the variable **Age When First Married** in the **Variable** box, and place the variable **RS Highest Degree** in the **Category Axis** box, and **Respondent's Sex** in the **Define Clusters By** box.
3. Click **OK**.

Q7. Give the five-number summary for men and women separately for each of the five categories of educational level attained. Draw a general conclusion about the effect of education and gender on the age at first marriage of the individuals of this sample. Notice that no generalization should be attempted at this level: we are only talking about the individuals of this sample and not about the general population.

_____
_____
_____
_____
_____
_____
_____
_____
_____

Conclusion:_____
_____
_____

LAB 6: MORE ON DESCRIPTIVE STATISTICS

In this lab, we will explore some of the difficulties arising when using SPSS. We will pursue with an exercise that consists in writing a report of a descriptive analysis.

Meaningless Numerical Results

SPSS does what you order it to do, even when it does not make sense. To see this, obtain the mean of any qualitative variable, such as **Marital Status** (you can do it through the **Descriptives** or **Frequencies** commands).

1. What is the mean for the variable **Martial Status**? Answer:_____
    Discuss with your peers: How appropriate is the Mean as a measure of central tendency for the variable **Marital**? Is there a better statistical procedure for describing the distribution of this variable? Which one is it? Using that procedure, describe the distribution of this variable.

2. Is the Mean appropriate to describe that variable? Give a reason for your answer:

    _____

    _____

3. Indicate a more appropriate statistical procedure to describe that variable:

    _____

    _____

4. Write down the numerical results obtained by this more appropriate procedure (copy the first three columns of the tables):

    _____

    _____

    _____

    _____

    _____

    _____

Mean When the Values of a Variable are Coded

When a quantitative variable is grouped into categories that are coded, statistical procedures will be performed on the codes, not on the value of the variable itself. To illustrate this, consider the variable **rincome91**. Look at how it has been coded.

5. Find the mean of the variable **Respondent's Income**: _____
6. Find the income range that corresponds to that mean: _____
    (consult the **Variables…** under **Utilities** to answer that question)

If you are asked to find the mean individual income, you will say it is somewhere in the range you gave in question 6 above. This procedure does not allow you to get the exact average income of individuals, because the actual income of individuals is not recorded in this data file. Only the category they belong to.

Important note: There are two more methodological problems with this mean:

1. The categories of income do not have an equal range: some span a range of $1000, some a range of $15,000 and some an open range (the category: $75,000 and more). These unequal ranges distort the value of the mean.
2. The category 22 (Refused to answer) is not labeled as a missing value, which means that it is counted in the computation of the mean, artificially inflating the answer.

Does this imply that we cannot use the mean of a coded variable? No. But we have to be careful in interpreting the answers. A mean computed in this way should not be taken as a precise measure of the average income, but as indicative of the *order of magnitude* of the mean income. It can also be used for comparisons: comparing the mean income of men and women, for instance, keeping in mind that the distortions in the mean are not identical across the categories: there could be more men than women in the Refused to answer category.

Exercise

Recode the variable **Respondent's Income** under the name **Respondent's Income Recoded** by replacing each code by the midpoint of the corresponding class interval (the Recode procedure was shown in Lab 3). For instance, the code 1, corresponding to income range $0 to $1000, would be replaced by 500. The category corresponding to the code 21 is problematic, as it is open. We can replace this code by 100,000 for the sake of this exercise; even if the answers are not precise, we will at least get a good order of magnitude. As for the category 22, which corresponds to 'Refused to answer', recode it as a system-missing value (the dialog box for recoding allows you to do that). Now compute the average income for the whole sample, then for specific categories: men vs. women, those who have a university degree vs. those who don't, etc. Such computations can be done with what you learned in Lab 3, using the **Explore...** command. Save the new data file that includes the recoded variable you have just created.

Doing a Comprehensive Analysis and Writing a Report

On the basis of the explanations given in Chapter 4, and in particular of the criteria given for a good analysis, answer the following question.

> Describe the sample in the **GSS93 subset** data file from the point of view of the family relationships of the individuals. Use all the variables that define the family, or that refer to the relationships within the family or to the opinions about such relationships. (Two to three pages, including tables and graphs.)

LAB 7: RANDOM SAMPLING

The purpose of this lab is to learn how to select cases in a file for more specific analysis, and how to select a random sample of cases for further analysis.

Random Number Seeds in SPSS

To select random samples, SPSS uses an electronic equivalent of the table of random numbers. The sequence of digits that are used as a starting point for selecting a sample is a *random number seed*. Thus, a random number seed is the electronic equivalent of a line or column in the table of random numbers. There are 2 billion random number seeds in SPSS, and when you ask SPSS to select a random sample, you can determine the number seed that will be used. If you do not specify the random number seed, SPSS will pick 2000000 as the default seed, that is, the seed it uses automatically. We will now select a random sample of 100 cases in the **GSS93 subset** data file.

Selecting a Random Sample

1. Open the **GSS93 subset** data file.
2. Determine the random number seed: Under the menu **Transform**, select **Random number seed**… .You get the box shown in Figure 7.1. Set the seed to 2000156. You may choose any other seed if you prefer, so as to get a sample that is not identical to the ones your colleagues are getting.

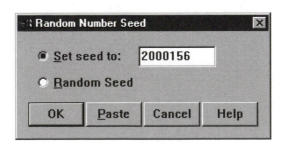

Lab 7 Figure 7.1

3. Click **Select Cases** … under **Data** menu. You get the dialog box shown in Figure 7.2.
4. Click on the button **Random sample of cases**, then double click on the **samples**… button which is right underneath it. You get the dialog box of Figure 7.3. You can see that you have a choice between selecting a percentage of cases, or an exact number of cases.
5. Click on the **Exactly** button, then write **100** in the first blank space, and **1500** in the second blank space, to indicate that you want to pick randomly 100 cases among the 1500 cases that are available in the **GSS93 subset** data file.

Lab 7 Figure 7.2

Lab 7 Figure 7.3

6. Click **Continue**. You get back to the first window, but now the sentence: **Exactly 100 cases among the first 1500** is written next to the button **Samples**....

7. Towards the button of the window, you can choose what to do with unselected cases. Click on the button for **Filtered**. This means that unselected cases will not be deleted. They will be seen on the screen, but a slash will be shown on their line number and they will not be taken into account in the computations.

8. Click **OK**.

Now look at the data file. How many cases are not crossed out? To answer that question, give a **Frequencies**... command to see the frequencies of the variable **gender**. Fill in Table 7.1 for the selected cases.

Table 7.1

	Count	Percentage
Males		
Females		
Total		

You can display a list of the selected cases by giving a **Frequencies...** command and choosing the variable **Respondent's ID**. You will get a list of the IDs of the selected cases. Any statistical procedure chosen will be performed on the 100 cases that have been selected. Give the **Frequencies...** command and indicate you want to compute the mean, median, max, min, and standard deviation for the variable **Respondent's Income Recoded** that you have created in Lab 5.

Save the resulting output on your floppy disk.

To de-select the cases and come back to the whole data file, do the following:

1. Choose **Select cases...** under the **Data** menu.
2. Select the button **Select all cases** and click **OK**.

Any procedure executed now will take into account all of the 1500 cases.

LAB 8: ACCESSING DATABASES ON THE INTERNET

The purpose of this lab is to learn how to explore and find information in some of the official websites of national or international organizations that are a source of primary data.

Recall from Chapter 7 that there are four types of documents you can find in an online statistical database:

- data sets of individual data (usually primary data);
- tables containing aggregate data;
- short descriptive reports highlighting the important statistical features of a situation;
- in-depth analytical report, containing data as support for the conclusions.

A given website may not include all four types, and if it does, access may not be permitted or free to everybody. Moreover, there are two ways of accessing given data: either by the table of links of the website, which leads us from general topics to increasingly detailed ones, or through the search function of the website.

For the exercises of this lab, you may choose three sites among the following. You will then try to find the same information in the three sites and see how they compare. The first site, Statistics Canada, includes in its home page a link called 'Links to other sites' that allows you to access directly the national statistical institutions listed here and many others as well.

National Institutions:

Statistics Canada	http://www.statcan.ca/
Australian Bureau of Statistics	http://www.abs.gov.au/
UK Office for National Statistics	http://www.statistics.gov.uk/
Central Statistics Office of Ireland	http://www.cso.ie/
China Statistical Information Network	http://www.stats.gov.cn/english/
Department of Statistics, Malaysia	http://www.statistics.gov.my/
Department of Statistics, Singapore	http://www.singstat.gov.sg/
Indonesian Bureau of Statistics	http://www.bps.go.id/
Statistical Center of Iran	http://www.sci.or.ir/
Statistics Bureau & Statistics Center (Japan)	http://www.stat.go.jp/

Statistics New Zealand	http://www.stats.govt.nz/
The Census and Statistics Dept. of Hong Kong	http://www.info.gov.hk/censtatd/
U.S. Census Bureau	http://www.census.gov/
International institutions	
UN/ECE Statistical Division	http://www.un.org/Depts/unsd/
International Monetary Fund	http://www.imf.org/
UNESCO Institute for Statistics	http://www.uis.unesco.org

Exercise 1. Exploring National Statistical Websites

Choose any three national statistical websites and compare how easy or difficult it is to find, through the table of links, the following information:

1. Demographic statistics, such as the total population by sex, as well as other demographic indicators such as the fertility rate and the rate of natural increase for the country.
2. Health indicators, such as life expectancy or the rate of occurrence of certain diseases, or statistics on medical services (such as number of doctors per 10,000 people)
3. Other social statistics concerning the structure of families, or crimes and the Justice system.
4. Through the search function, find data and descriptive or analytical documents with the search keyword 'marriages'.
5. Compare the ease of access of these documents to the ones you can get through international bodies such as the UN system or the UNESCO statistics, whose website address is given above.

Exercise 2: Specific Search in the Statistics Canada Website

1. Go to the StatCan website (http://www.statcan.ca/). You may want to do such a detailed search on some other national statistical website, but you may find that the data is organized differently. You will still be able to find comparable information most of the time.
2. Click on one of the two buttons, **English** or **French**. Take a minute to see what is on the page. List all the links that are found on the page.
3. Click on **The Daily**.
4. List the major releases of the day. List the other releases of the day. The major releases are not published everyday. But links to other issues of the daily are offered.
5. Pick one of these releases. Summarize in your own words, in half a page, the information found in this report.

6. Go to the **Census** link. You will find information on the 2001 Census, including
    the planned released dates of statistical reports from that Census, together with
    statistical data relating to the 1996 Census. Select the **Nation Series** in the 1996
    Census, and then the Age and Sex tables. An additional link allows you to view
    the tables in HTML format.

    *   Write down the number of people recorded in the 1996 Census in Canada.
    *   Find the percentage of people living in Canada who are under 15 years of age.
    *   Find the number and percentage of men and of women in Canada.
    *   Find the percentage of men and of women over 75 years old.
    *   Explore the other tables.

7. Go to the **Ethnic Origin** tables, in the **Nation series** of the **Census**.

    *   How many people in Quebec said they had a *single* Italian origin?
    *   How many said they had an Italian origin AND some other origin?
    *   Is this data taken on the whole population or just on 20% of it?

8. Go to the **Mother Tongue** tables

    *   What are the categories used to describe the variable Mother tongue?
    *   What is the percentage of people whose mother tongue includes French?

9. Go back to the Home page of StatCan, and perform a search with the keyword
    Marriages. This search turned up 25 documents when we did it. The report
    reproduced as an example in Chapter 4 is one of those. Explore the other reports
    that turned up in this search.

LAB 9: CREATING A DATA FILE

The purpose of this lab is to learn how to create a data file starting from a questionnaire, how to include all the required specifications of the variables that are created, how to enter data, and how to save and print a data file and an output file.

1. Opening a New Data File

Recall that an electronic data file consists of information that has been collected, then entered with the help of some statistical software and organized in a way that facilitates its statistical analysis. Each column represents one **variable**, and each line represents one **case**.

When you open SPSS, you get a window that offers several options, including:

O   Type in data

and

O   Open an existing data file

Click on the round button preceding the words **Type in data**: you get a blank data file. Sometimes, the program is set so that the **Type in data** option is selected automatically, and it opens directly on the blank data sheet.

The columns of the data sheet correspond to variables, and the lines to cases. Before you type the data, you must specify the characteristics of each of the variables you want to introduce. It is always preferable to prepare a blank data matrix before entering any data, and to print the File Info to check if the variables have been defined properly. We can always modify the characteristics of a variable afterwards, and even add new variables, but it is preferable to start with a good blank matrix where all the variables have been defined. The word 'matrix' is used to designate a blank data file where all the variables have been defined. Once the data is typed in, we usually call it a data file.

To construct the SPSS matrix, go to the Variable View of the Data Editor (Figure 9.1).

On the top of your window, the name of the data file is written. In this example, the word **Untitled** appears because we have not named the data file yet.

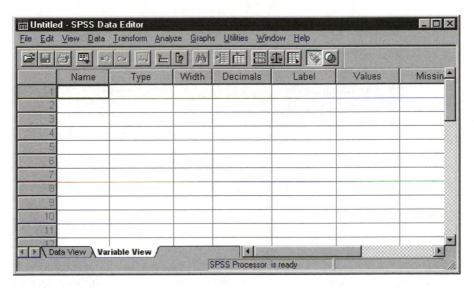

Lab 9 Figure 9.1

Look carefully at the various columns in the Variable View of the data editor. Each line is a variable, and each column concerns one of the characteristics of the various variables. The various columns are:

Name, Type, Width, Decimals, Labels, Values, Missing, Columns, Align, Measure. As we explain how to fill in the details that relate to a variable, we will illustrate the procedure with the variable Respondent's Sex. We will code the variable Respondent's Sex as:

1   Men
2   Women

and give it the name 'sex'.

1.   The first column to fill in is the variable **Name** column. This must be a short name, with at most eight characters and no spaces. Type the name 'sex'.

2.   The second column is the **Type**. Click on the right side of the cell; you get the dialog box shown in Figure 9.2. Here you indicate whether the data is numerical, or a sign such as a dot or a comma, or a currency, or a date, or a string variable.

   A **string** variable is a variable made out of letters and digits, and it is used when you want to type in a proper name, such as Peter or Mary. You cannot do many statistics with string variables, except count how often a given string shows up in your data. Very few statistical procedures apply to string variables.

   A **numeric** variable is used when you want to type in your data as numbers. If these numbers refer to actual numerical values you have to specify how many

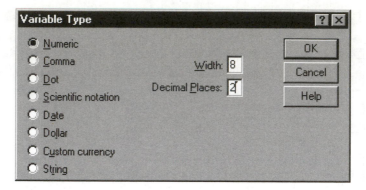

Lab 9 Figure 9.2

spaces and decimals you need. For example, if you want to record the height of
the respondent measured in centimeters, you need 5 digits with one decimal
place, so as to be able to write numbers like 172.3 cm (the dot uses a space). If
the numbers refer to categories (for example: 1 = Female; 2 = Male), you only
need a width of 1 digit, and no decimal places. You will need two digits if you
have more than ten categories.

3. The third column is **Width**. It was set in the previous step to 8 digits, but it could
    be changed directly in that column to a larger or smaller number of digits.

4. The fourth column is the number of decimals, which was also already set, and
    which can also be changed directly in this column.

5. The fifth column is the variable **Label**. Type: **Respondent's Sex**. This is the name
    that will appear on your output tables, so it must be chosen carefully to convey
    the meaning of the variable. For this variable this is not an issue, but there are
    other cases where you have to take the time to choose the variable label to be
    both precise and concise.

6. The sixth column is the **Value Labels**. Click on the right side of the cell. The
    dialog box shown in Figure 9.3 should pop up.

Lab 9 Figure 9.3

Type:

> Value: **1**
> Value Label: **Male**

Then click **Add**, and type:

> Value: **2**
> Value Label: **Female**

Now click **Add**, then **OK**.

7. The seventh column is where you indicate how to code the Missing Values. We could leave it set to the default setting: None, as it is expected that the sex of the respondent is known. But we can think of situations where the sex of the respondent is not known (for instance if it is determined by the voice in a telephone communication). In such cases you may want to enter a missing value. We will click on the right side of the cell, and get the dialog box shown in Figure 9.4. The window gives you four choices. Either you have no missing values, or you have up to three distinct missing values, or you have a whole range of missing values, or you have a whole range of missing values plus one more separate missing value. The word discrete means separate and distinct; it is the opposite of continuous. You can have up to three discrete missing values. If you decide that, say,

> 7 stands for NA (No Answer)
> 8 for DK (Don't Know), and
> 9 for NAP (Not Applicable),

then your should enter the values 7, 8 and 9 in the Discrete missing values boxes, but you should also go back to the Value Labels column and enter the three corresponding value labels (NA, DK, and NAP).

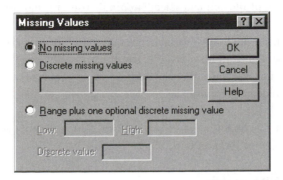

Lab 9 Figure 9.4

Click on Discrete Missing Values, type **9** in the first box, and click **OK**. Now we have to indicate what 9 stands for. Go back to the Value Labels columns, click on the right side of the cell, and add:

Value: **9**
Value Label: **Not known**

and click **Add**. The box should look as in Figure 9.5 before you click **OK** and close it. It is important to designate Missing values properly otherwise SPSS will count them in the statistics, which could be misleading.

Lab 9 Figure 9.5

8. The eighth column determines how wide the column looks on the screen. Leave it at 8 digits.
9. The ninth column determines whether the data for this variable will be right aligned, centered or left aligned on the screen. Leave it as it is.
10. The last column is very important. It determines the level of measurement of the variable: **scale**, **ordinal** or **nominal**. For the variable **Sex**, we will click on the right side of the cell and select **Nominal**. You may have noticed that some of the SPSS examples label qualitative variables such as sex as 'ordinal'. This is done when a qualitative variable has only two categories (that is, when it is dichotomous) and it allows us to perform certain advanced quantitative techniques on such variables. For the purpose of this course, we will not need to use ordinal measurement scales to code qualitative variables.

The end result of this operation is shown in Figure 9.6, which shows also a second variable, called **Age of Respondent**.

Checking the SPSS matrix

After you have defined all your variables, you may want to check that you have made no mistake. The best way to do it is to enter a **File Info** command (**Utilities → File**

Lab 9 Figure 9.6

Info) and examine the output. The output for the two variables of this illustration should look as follows:

```
        List of variables on the working file

Name                                                    Position
SEX                      Respondent's Sex                  1
Measurement Level:       Nominal
Column Width:            8 Alignment: Right
Print Format:            F8
Write Format:            F8
Missing Values:          9

   Value                 Label
   1                     Male
   2                     Female
   9            M        Not Known
   AGE                   Age of Respondent               2
Measurement Level:       Scale
Column Width:            8 Alignment: Right
Print Format:            F8
Write Format:            F8
Missing Values:          999
   Value                 Label
   999          M        No Answer
```

All the definitions of the variables appear in this output. You can check that the variable name and label are correct, and that the value labels and codes are what you want them to be. You must also check the format and the missing values, as well as the codes for the missing values.

Only when you have entered all your variables and have checked that they are correct can you start typing your data in the Data View window of SPSS. It does not matter whether you type your data one line at a time or one column at a time. After you have typed the data in one of the cells, if you press the **Tab** key on your keyboard you will move to the next cell on the same line, but if you press the **Enter** key you will move to the cell below in the same column.

When you enter the data, you must enter the values, not the labels. Thus, you would type

1   for Male,   and
2   for Female.

For a quantitative variable such as Age of Respondent, you would enter the age itself. There are no value labels for this variable, except for the missing value (999) which is labeled 'No Answer'.

Finally, when you type in your data in the Data View window, you may have the Value Label command, under View, turned either on or off. If it is not checked, you will see '1' on the screen when you type '1'. If it is checked, you will see 'Male' when you type '1' (or whatever the value label is for the variable you are typing).

Practical Exercise

Create an SPSS data file to enter data collected with the help of the following questionnaire. The blank questionnaire is given, followed by the data collected with the help of that questionnaire. Create the SPSS blank matrix from the questionnaire (do not forget to include missing values whenever relevant), print the File Info to check that the variables have been created correctly, then type in the data given below and save the data matrix on your disk. To make sure you have entered the data correctly, produce the frequency table for each of these variables and examine it to see if it corresponds to the data given.

Questionnaire

Questionnaire number: _____

1. What is your sex? M_____ F_____
2. What is your age today? _____ (in years)
3. What is your height in inches? _____
4. What is your father's height in inches? _____
5. What is your mother's height in inches? _____
6. What is your hair color? (put a check mark in front of the color that is closest to yours)

    ❏ Black
    ❏ Brown
    ❏ Blonde
    ❏ Red
    ❏ Light brown/fair

7. What is the color of your eyes?

    ❏ Black
    ❏ Brown
    ❏ Blue
    ❏ Green
    ❏ Other

8. Have you ever dyed your hair? Yes _____ No _____
9. Are you left-handed, right-handed, or ambidextrous? _____
10. Do you wear contact lenses? Yes _____ No _____

Answers to the above questions given by 10 subjects:

Table 9.1

Q1	Q2	Q3	Q4	Q5	Q6	Q7	Q8	Q9	Q10
M	20	68	66	69	Black	Green	No	Left	No
F	18	67	67	68	Brown	Blue	No	Right	Yes
F	22	68	66	69	Blonde	Black	Yes	Right	Yes
F	22	67		72	Black	Blue	Yes	Ambid.	Yes
M	20	68	69	67	Brown	Brown	No	Right	No
M	28	66	68	65	Brown	Blue	DK	Right	Yes
F	18	67	67	69	Brown	Brown	Yes	Right	No
M	17	69	70	66	Brown	Blue		Right	No
M	22	68	68	66	Blonde	Blue	No	Right	No
F	30	71	71	68	Brown	Brown	Yes	Ambid.	No

LAB 10: CROSS-TABULATIONS (TWO-WAY TABLES)

The purpose of this lab is to learn to test statistical association for categorical variables, that is, to learn how to produce and interpret two-way tables with or without the various cell percentages, and to practice the interpretation of such tables.

What is the Question We Are Trying to Answer?

We want to know whether there is a statistical association between two nominal variables. Recall that the meaning of 'statistical association' is the following: We say that there is a statistical association between two nominal variables if the individuals who score one way on one variable *tend* to score in a specific way, similar to each other, on the other variable. Refer to your notes on statistical association for more details.

Illustration with SPSS

SPSS can compute the **two-way table** (also called the **contingency table**) of two nominal variables. The operation itself is called **cross-tabulation**, and the two-way table is the *result* of the operation. SPSS will also do it for two quantitative variables even if they are measured by an ordinal or numerical scale. However, such tables are useful only if you have a small number of values for each of the variables.

Select **Analyze → Descriptive Statistics → Crosstabs**.

You get the dialog box shown in Figure 10.1. Notice that we have already placed the variable **degree2** in the box for rows, and **income4** in the box for columns. This is because we wanted to see whether there is a statistical association between education and income. Education is measured in this analysis by the variable **degree2** (college degree; 2 categories: yes or no), and income by the variable **income4** (Total family income, 4 categories).

Notice also that the variable **income4** is a quantitative variable, but that the values have been grouped into 4 categories. It makes sense therefore to use a procedure that works for categories.

Look carefully at the various items to be specified in the dialog box of Figure 10.1. You can get a *clustered bar chart*, and you can also specify what will appear in the cells of the table. Click on **Cells...** . You get the dialog box illustrated in Figure 10.2.

Lab 10 Figure 10.1

Lab 10 Figure 10.2

In the **Counts** box, click on **Observed**. Do not click on any of the percentages for the time being. The other commands relate to inferential statistics and we will not discuss them here.

Click **Continue** then **OK**. You get two tables. The first one gives you technical information about the valid answers: how many units we have in the sample, how many answered both questions, and how many answers are missing. This table is important for you, but is should not be printed in a report. It is the second table that is more interesting: this is the contingency table we have already learned to read. Here it is (Table 10.1).

Table 10.1 **College Degree *Total Family Income Cross-tabulation**

| College Degree | | **Total Family Income** | | | | |
		24,999 or less	**25,000 to 39,999**	**40,000 to 59,999**	**60,000 or more**	**Total**
No College degree	Count	536	224	160	229	1149
College degree	Count	48	76	70	153	347
Total	**Count**	**584**	**300**	**230**	**382**	**1496**

This table gives the number of individuals in every cell, but it is not easy to interpret, because the two categories **College Degree** and **No College Degree** do not contain the same number of units. If you just look at the category **$60,000 or more**, for instance, you may have the (wrong) impression that individuals without a college degree are more represented in it than individuals with a college degree. This is so because the former is more numerous in the sample as a whole (1149 individuals) than the latter (347 individuals). In order to say whether people with a college degree are *more likely* to be in the high-income category than others, we would have to look at percentages. Let us find the row percentages, that is, the percentages of people *within* the **College Degree** categories.

Go back to the **Crosstabs** command to modify it. Click on the **Cells…** button. In the **Percentages** box, click now on **row**. You get Table 10.2.

Table 10.2 **College Degree *Total Family Income Cross-tabulation**

| College Degree | | **Total Family Income** | | | | |
		24,999 or less	**25,000 to 39,999**	**40,000 to 59,999**	**60,000 or more**	**Total**
No College degree	Count	536	224	160	229	1149
	% within College Degree	46.6%	19.5%	13.9%	19.9%	100.0%
College degree	Count	48	76	70	153	347
	% within College Degree	13.8%	21.9%	20.2%	44.1%	100.0%
Total	**Count**	**584**	**300**	**230**	**382**	**1496**
	% within College Degree	39.0%	20.1%	15.4%	25.5%	100.0%

Analysis

A careful analysis of the numbers above is in order. Consider first the category of people who make $60,000 a year or more. Notice that a majority of them do not have a college degree, and only a minority have a college degree. Should we conclude therefore that if you do *not* have a college degree, you have a *better* chance of making $60,000 or more? Compare the similar percentages in the category of people who make less than $25,000 a year.

The percentages tell us the following: looking at the various cells in the row **No College Degree**, we can conclude that individuals in this sample who do not have a college degree are *more likely* to be in the low income categories: close to 47% of them are in lowest income category, and only 20% in the highest income category. The percentages for the individuals of this sample with a college degree are the opposite: 14% in the lowest income category and 44% in the highest income category.

Conclusion

There is a statistical association between having a college degree and having a high income: Individuals in this sample who have a college degree are more likely to be in the high income categories than individuals who do not have one.

Notice that we did not generalize our conclusion to the whole population: we are talking about this sample of 1500 individuals. We will learn in the chapter on inference how to make generalizations to the whole population from which the sample is taken.

Which Variable Should We Place in the Rows?

When doing a cross-tabulation, we have to decide which variable will be placed in the rows of the table and which will be placed in the columns. Basically, it does not make a great deal of difference, and the same analysis can be performed in both cases. However, it is customary to place the independent variable in the rows, and the dependent variable in the columns. The dependent variable is the one you wish to explain. In the example above, if we want to explain the income level by the educational level, we should place the variable **degree2**, which measures the level of education, in the rows, and the variable **income4** in the columns, and we should ask to get the row percentages. The output is easier to read that way. However, we could still perform the same analysis and come to the same conclusions if the rows and columns were interchanged. Simply stated, for an easier reading of the table: *Place the independent variable in the rows, and have SPSS display the row percentages.*

Exercises

Analyze the statistical association between the pairs of concepts indicated below. For each pair, select the appropriate variables from the **GSS93 subset** data file, produce the contingency table, and analyze the table. Recall that at this point, we are not generalizing to a whole population. Your conclusions will simply describe the association between the two variables *for the sample of 1500 individuals included in this data set.* Recall also that statistical association is not the same as a causal relationship.

Your conclusions should be expressed in the following format:

> *This table shows that there is (or: there is not) an association between*
> *... ... and ... . People who ... . tend to (or: do not tend to,* if there is
> no association).... *more than those who do not* .... (At this point,
> throw in some of the percentages that support your conclusions.)

(a) gender and opinion about the death penalty for murder;
(b) gender and opinion about the legalization of marijuana;
(c) gender and voting in the 1992 election (vote92);
(d) opinion about the death penalty and opinion about gun permits;
(e) religious preference and opinion about the death penalty for murder.

Write your answers on a separate sheet, or type them on the computer and print them.

LAB 11: COMPARING THE MEANS OF VARIOUS SUBGROUPS

The purpose of this lab is to learn how to test statistical association for one categorical variable and one numerical scale variable and to interpret the the SPSS outputs.

We have seen in Lab 3 that the **Explore** command (**Analyze → Descriptive Statistics → Explore**) can be used to compare the mean of a quantitative variable for various subgroups of the sample. For instance, consider the variable **Age when First Married** in the **GSS93 subset** data file. You may want to know the average age at which men marry for the first time, and compare it with the average age at which women marry for the first time. Or you may want to know whether people who have a college degree tend to marry later then those who do not have one. The **Explore** command allows you to do that.

But there is another command that allows you to do such comparisons in a more detailed way. For instance, you may combine the groups, and look at the average age when first married for:

- men with a college degree;
- women with a college degree;
- men without a college degree;
- women without a college degree;

The procedure that allows you to do such an analysis is the **Compare Means** command, which is found under the **Statistics** menu.

Keep in mind which level of measurement is needed for such an analysis. This procedure, like the **Explore** command, is applied when you want to find the mean of a *quantitative variable, measured by a numerical scale* (which is then the *dependent* variable), for various subgroups defined by some other categorical variable (the *independent* variable; see Table 11.1). Since we are talking about subgroups of the sample, the independent variable must divide your population into subgroups that you wish to compare. Therefore, it must be measured at the nominal or ordinal levels. It could also be measured at the numerical scale level, provided the values are grouped into a small number of categories. Here are some examples where the procedure would apply.

Table 11.1

Dependent Variables	Independent Variables
Age when First Married (**agewed**)	Respondent's sex (**sex**)
Respondent's Age (**age**)	College degree (**degree2**)
Respondent's income (**rincome91**)	Age Categories (**agecat4**)
Highest Year of School Completed (**educ**)	Marital status (**marital**)
etc.	Labor Force Status (**wrkstat**).
	etc.

You may have noticed that the independent variables are either qualitative (such as **Respondent's Sex**) or are quantitative but grouped into a small number of categories (such as **Age Categories**).

The Procedure Compare Means

Here is how the procedure applies.

1. Determine which variable you want to examine (the dependent variable) and which subgroups you want to consider (such subgroups should be defined by some independent variable). For instance, we may want to see whether the men of this sample tend to marry at a different age than the women of the sample.
2. Select **Compare Means** under the **Statistics** menu, then select **Means**. You get the dialog box of Figure 11.1 We have placed the dependent and independent variables in their respective boxes.
3. Click **OK**. You should get the output shown in Table 11.2.

Lab 11 Figure 11.1

Table 11.2 **Report: Age When First Married**

College Degree	Mean	N	Std. Deviation
No College degree	22.10	936	4.75
College degree	25.26	265	5.26
Total	22.79	1201	5.03

You can see from Table 11.2 that the individuals of this sample who have a college degree tend to marry more than 3 years later then those who do not have a degree. Moreover, their age at first marriage tends to be slightly more dispersed than those who do not have a college degree. You know that by looking at the standard deviation, which is a measure of dispersion. A standard deviation of 5.26 indicates that the distribution of their ages is slightly more dispersed than those who do not have a college degree (st.dev. of 4.75).

Notice that you did not change any of the options that are built-in by default into this procedure. By clicking on the **Options** button, you could have more information shown in the output (Table 11.1), but you do not need to do that for the time being.

Combining Variables

You can also use the **Compare Means** procedure with more than one independent variable. To understand how to interpret the results, let us go through the following steps.

1. Go back to the **Compare Means** procedure shown above, and place the dependent and independent variables exactly as above. Notice that in the middle of the window, there is the phrase: **Layer 1 of 1** and there is a button called **Next**.
2. Click on the **Next** button. You will notice that the sentence now reads **Layer 2 of 2**. Place the variable **Respondent's Sex** in the independent variable box, and click **OK**. You get the output shown in Table 11.3.

Table 11.3 **Report: Age When First Married**

College Degree	Respondent's Sex	Mean	N	Std. Deviation
No College degree	Male	23.56	356	4.72
	Female	21.20	580	4.54
	Total	22.10	936	4.75
College degree	Male	25.75	136	4.91
	Female	24.74	129	5.57
	Total	25.26	265	5.26
Total	Male	24.16	492	4.87
	Female	21.84	709	4.93
	Total	22.79	1201	5.03

You see now that the group of individuals having a college degree is broken down into Male and Female having a college degree, and the average age at first wedding is computed separately for each subgroup. Similarly, the group of individuals not having a college degree is broken down by sex, and the average age at first marriage is given for each subgroup.

The following exercise will get you acquainted with the interpretation of the Compare Means Procedure.

Exercise 1

Fill out the blanks in the following sentences:

> The men of this sample who have a College degree marry for the first time on the average at: _____ years.
> The women of this sample who have a College degree marry for the first time on the average at: _____ years.
> The people of this sample who have a College degree marry for the first time on the average at _____ years.
> The men of this sample who do not have a College degree marry for the first time on the average at _____ years.

Comparing the Effect of the Two Variables

The table given above allows us to compare the effect of each of the two variables **Sex** and **Education** on the ages at first marriage. Indeed, the table gives the average age at first marriage for those have a college degree (males and females) and the average age for those not having one. We see that the averages are, respectively, 25.26 and 22.10. This means that those who get a college degree get married on the average 3.16 years later than those who do not get one. On the other hand, we can find the average age at first marriage for males and females separately, not taking into account whether they have a college degree or not. We see here that males tend to get married on the average 2.32 years later then females do. We can therefore conclude that getting a college degree has an effect which is more important than that of gender on delaying the age of marriage. Gender tends to delay the age of marriage by little over 2 years, while getting a college degree tends to delay the age of marriage by a little over 3 years.

The assertions above are examples of the kind of conclusions that can be reached by using the **Compare Means** procedure.

Exercise 2

Select a quantitative variable of your choice, and analyze its means by breaking the sample down into several groups using two different independent variables. (Example: you could perform an analysis of the variable **Respondent's Income** similar to the one above, looking at the combined effect of sex and educational level. Recall however that what you get in **Respondent's Income** is a code, not a dollar value. You may want to use the variable **Respondent's Income Recoded** that was created and saved in Lab 4.)

      Dependent variable chosen:_____

      First independent variable chosen:_____

      Second independent variable chosen: _____

      Conclusions (Supported by numbers and percentages, and formulated in full sentences):

Exercise 3

Interchange the role of the first and second independent variables. What difference does it makes for the output? Is the output different just in form, or do you get different information?

LAB 12: CORRELATION AND REGRESSION

The purpose of this lab is to learn how to test statistical association for two quantitative variables, measured at the numerical scale level, and to produce and interpret coefficients of correlation and the regression line and their equations.

The procedures that follow apply to *quantitative* variables only, those that are measured at the numerical scale level of measurement (also called interval or ratio scales). To perform a correlation analysis and produce the relevant diagrams, go through the following steps.

1. Open the SPSS file entitled **Road Construction bids**. You do that by selecting **More files...** when you first open SPSS; you get a list of files and you can select this one.

2. Take a few minutes to read the names and labels of the variables involved. This file deals with the costs and the number of days of work of various projects undertaken by the **DOT**, which stands for the **'Department of Transport'** of some city.

3. We will study the relationship between two quantitative variables: the estimated cost of a project, and the real cost of the same project. The estimated cost is given by the variable **dotest**, which stands for 'DOT Engineer's Estimate of Construction Cost', and the real cost is given by the variable **cost**, which stands for 'Construction cost'. We want to know whether the estimates of the DOT Engineers are accurate.

4. Select **Analyze**
        Correlate
              Bivariate...

   You get the dialog box shown in Figure 12.1.

5. Place the two variables **Construction cost** and **DOT Engineer's Estimate of Construction cost** into the variables: box).

6. Click **OK**. You should get Table 12.1.

You can see that for the two variables under consideration, the coefficients of correlation is 0.987, which is a very strong correlation. This means that, in general, the estimated costs are a good predictor of the real costs (but not necessarily *equal* to the real cost: you may have to multiply the predicted cost by a factor to get the real cost).

Lab 12 Figure 12.1

Table 12.1 **Correlations**

		Contract Cost	DOT Engineer's Estimate of Construction Cost
Contract Cost	Pearson Correlation	1.000	.987
	Sig. (2-tailed)	.	.000
	N	235	235
DOT Engineer's Estimate of Construction Cost	Pearson Correlation	.987	1.000
	Sig. (2-tailed)	.000	.
	N	235	235

** Correlation is significant at the 0.01 level (2-tailed).

Notice that for every pair of variables, the table includes the coefficient of correlation, also called the Pearson Correlation coefficient, as well as the level of significance, denoted by Sig (2-tailed), which has not been studied yet. We will learn about its meaning later on. It also includes the number N of cases taken into account.

Notice also that some of this information is redundant and could be omitted. The correlation between a variable and itself is always 1. So the corresponding cells could be omitted. Moreover, interchanging the order of two variables does not affect the correlation between them, which is why the cell at the top right is identical to the one at the bottom left. One of them could be omitted. The conclusion is that Table 12.2 carries exactly the same information as Table 12.1. The options on SPSS are sometimes set so as to give the correlation table in the format of Table 12.2.

Table 12.2 **Correlations**

		Contract Cost	DOT Engineer's Estimate of Construction Cost
Contract Cost	Pearson Correlation		.987
	Sig. (2-tailed)		.000
	N		235
DOT Engineer's Estimate of Construction Cost	Pearson Correlation		
	Sig. (2-tailed)		
	N		

** Correlation is significant at the 0.01 level (2-tailed).

The following steps tell you how to represent this situation graphically.

7. Select
   Graphs
      Interactive
         Scatterplot.

8. In the dialog box you get, slide the variable **dotest** on the horizontal axis, and the variable **cost** on the vertical axis. The dialog box should look as in Figure 12.2.

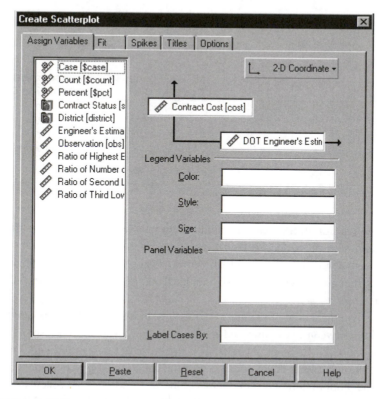

Lab 12 Figure 12.2

9. Now click on the tab called **Fit** on the top of this dialog box. You get a new dialog box shown in Figure 12.3. Select the options as shown in Figure 12.3, that is, make sure that the **Regression** option is chosen and that there is no check mark facing the word **Mean**.

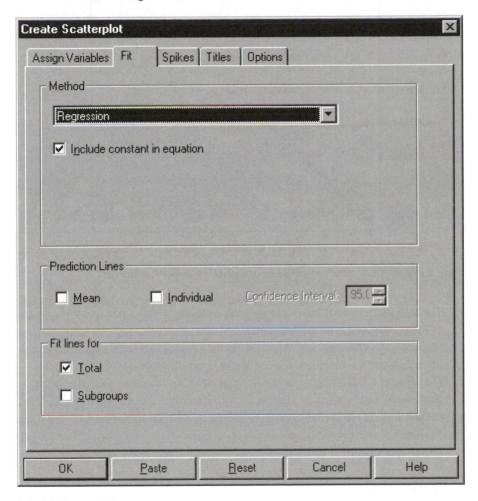

Lab 12 Figure 12.3

10. Then click on the tab called **Options**. You are offered a series of choices that relate to the visual aspect of the chart. Click any option. Figure 12.4 results from choosing 'Classic'.

11. Click **OK**. You get the diagram shown in Figure 12.4.

Figure 12.4 shows the regression line illustrating the relationship between the two variables, as well as the mathematical equation of that line. This is the visual representation of the relationship between the two variables. The conclusions that we can draw from that graph are the following:

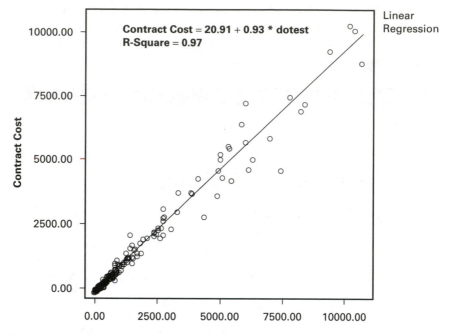

Lab 12 Figure 12.4   DOT engineer's estimate of construction cost

1. There is a strong correlation between the estimated cost of constructions by the DOT's engineers, and the real conctruction costs of the various projects. The correlation coefficient is 0.987, given in the previous table. Therefore, the estimated cost of a project is a very good predictor of the real cost.

2. However, the DOT engineers tend to overestimate the costs slightly. Based on past experience, we can predict that, on the average, the real cost of a contract will be slightly lower than the one estimated by the engineers by a factor of 0.93. We see that in the equation relating the two variables, which is printed on the top part of the diagram.

3. We also see that the prediction is much more accurate for small contracts than for big contracts.

4. For a given project, if we know the engineer's estimate of the cost, we can estimate the real cost in two different ways: graphically as shown in Figure 12.5, or using the regression equation. If the DOT's engineer's estimate is $5000, for instance, the regression equation tells us that the real cost should be:

$$20.91 + 0.93 \ (5000) = 20.91 + 4650 = \$4671 \ \text{approximately.}$$

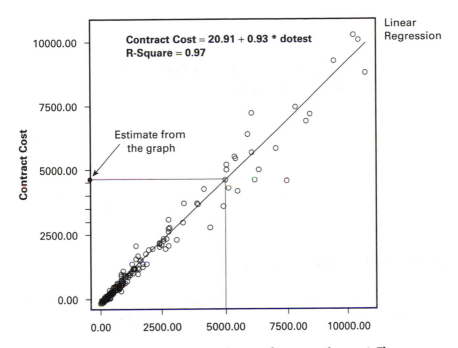

Lab 12 Figure 12.5   DOT engineer's estimate of construction cost. The actual contract cost corresponding to a DOT engineer's estimate of $5000 has been estimated manually to be roughly $4500, as shown in this figure

Exercise

Open the SPSS file **World95**, and examine whether there are correlations between the following variables. Express the conclusions as was done above.

- Average female life expectancy
- People who read (%)
- Infant mortality (deaths per 1000 live births)
- Daily calorie intake
- Birth rate per 1000 people

LAB 13: CONFIDENCE INTERVALS

The purpose of this lab is to learn how to produce and interpret the outputs for estimation.

Confidence intervals for the mean

SPSS can compute the confidence intervals for the mean of a quantitative variable. This is done through the **Explore** command, seen in Lab 3. Here is how it is done.

1. Open the **Employee Data** file.
2. Click on **Analyze,** then **Descriptive Statistics** then **Explore....**
3. In the box you get, you can select the variables you want to analyze. Select a quantitative variable, let us say **Months Since Hire,** nicknamed **jobtime,** and place it in the **Dependent variables** box.
4. You also have a button called **Statistics** that allows you to determine what you want to see calculated. Click on it. You get the dialog box shown in Figure 13.1. The word **Descriptives** is checked, and the confidence level proposed is 95%. You can change it to 99% or 90% if you prefer.
5. Click **Continue,** then **OK** in the **Explore...** dialog box.

You will get Table 13.1 in the Output window (some of the bottom rows have been deleted because we do not need them now).

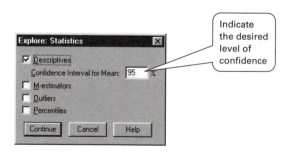

Lab 13 Figure 13.1

Table 13.1 **Descriptives**

				Statistic	Std. Error
Months Since Hire	Mean			81.11	.46
	95% Confidence Interval for Mean	Lower Bound		80.20	
		Upper Bound		82.02	
	5% Trimmed Mean			81.12	
	Median			81.00	
	Variance			101.223	
	Std. Deviation			10.06	
	Minimum			63	

Examine the meaning of every number in that table:

- The mean for that sample is 81.11 months of work since the employee has been hired.
- The confidence interval is given in the two shaded cells: from 80.20 to 82.02 months. What this means is that if the units in this file were a representative sample of a larger population of employees, you would be *estimating* the mean time since hire for that larger population to be somewhere between 80.20 months and 82.02 months (you can convert these decimals to days or weeks if you wish).
- The standard error is the standard deviation divided by the square root of n, a quantity which is used in the computation of the margin of error. But you do not need to use it, since the confidence interval has been computed by the SPSS program.
- The remaining measures have been seen in the chapter on descriptive statistics.

These notions have been explained in Chapter 9 and you should review that chapter in order to interpret correctly the results given by SPSS.

Exercises with SPSS

1. Write a confidence statement as a full sentence for the following situations, assuming that the data file is a representative sample of some population. The statements you write should have the same format as the ones in Chapter 9 or those provided in the exercise section of that chapter. Estimate the parameter of the variable **jobtime** in the **Employee Data** file, at a 95% confidence level:

   _____

   _____

   _____

   Estimate it at a 90% confidence level:

   _____

   _____

   _____

What is the difference in the estimated interval between the two statements? Explain why there is such a difference.

_____

_____

_____

2. Write similar statements for the variable **Salary** in the **University of Florida** data file (use a separate sheet).

Graphical Representation of Confidence Intervals for the Mean

Confidence intervals can be represented graphically as follows.

1. Under **Graphs**, choose **Error Bar**... .
2. In the dialog you get, select **Simple** and **Summaries of Separate variables**, then click **Define**.
3. Now choose a quantitative variable, and place it in the Errors bars box. Underneath, you will see a box where it is indicated **Bars Represent**: and the text in the space states: **Confidence interval for the mean.** And in this box, you will see a space where you want to indicate the desired confidence level (95%, or 90% or 99%).
4. Click **OK**. You will get a graph that represents the confidence interval, illustrated in Figure 13.2.

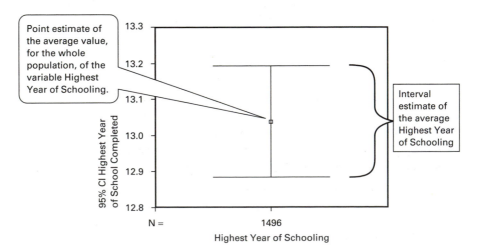

Lab 13 Figure 13.2

Confidence Intervals for a Proportion or Percentage

Consider the two variables:

> **cappun** Favor or Oppose Capital Punishment, and
> **attsprts** Attended Sports Event in Last Yr

which are found in the **GSS93 subset** data file.

These two variables are qualitative variables, and they are measured at the nominal level. Therefore, we cannot compute the mean, but we can compute the *proportion* (or percentage) of people in the sample who favor capital punishment, or who attended a sports event last year. Assuming the sample in this data file has been chosen at random, we can then estimate the corresponding percentages in the American population. But if there is a high percentage of missing data, the reliability of such estimates is questionable.

SPSS will not provide you with an estimate of the required percentages. SPSS will give you the required percentages in the sample, by using the Frequencies command. You can determine the margin of error at the 95% confidence level by using the formula given in the section on estimation. Table 13.2 gives the approximate value of the margins of error for various sample sizes and various values of the percentage that is found in the sample, at the 95% confidence level.

Table 13.2

Population Percentage	Sample size						
	100	**200**	**400**	**500**	**800**	**1000**	**1500**
Near 10	7	5	4	3	3	3	2
Near 20	9	6	5	4	3	3	3
Near 30	10	7	5	5	4	3	3
Near 40	10	7	5	5	4	4	3
Near 50	10	7	5	5	4	4	3
Near 60	10	7	5	5	4	4	3
Near 70	10	7	5	5	4	3	3
Near 80	9	6	5	4	3	3	3
Near 90	7	5	4	3	3	3	2

Margins of error for the estimation of a percentage, at the 95% confidence level.

Example

Get the frequencies for the variable **Favor or Oppose Death Penalty for Murder**. Look at the valid percent. You find that 77.4% of the valid answers are in favor of the death penalty for murder. Assuming that this sample is representative, you want to estimate the percentage of people who are likely to favor the death penalty for murder. In this case the percentage is close to 80%, and the number of valid answers is 1388, which is close to 1500. The table gives us a margin to error of 3%. Thus, the confidence statement becomes:

> On the basis of that sample, we can estimate that the percentage of Americans who are in favor of the death penalty for murder is somewhere between 74.4% and 80.4%, at a 95% level of confidence.

Exercise

3. Consider the two variables:

    letdie1 Allow Incurable Patients to Die, and
    Scitest4 Humans Evolved From Animals

    which are found in the **GSS93 subset** data file.

    Assuming the sample in this data file has been chosen at random, make an estimate of the percentage of adults in American society who believe that patients who suffer from an incurable disease should be allowed to die. Make also an estimate of those who believe that Evolution theory (which asserts that humans evolved from animals) is either *probably true* or *definitely true* (lump the categories together by adding their percentages).
    Considering the percentage of missing data, write a comment on the reliability of such estimates.

LAB 14: t-TESTS

The purpose of this lab is to learn to perform simple t-tests and to interpret their outputs. We will see here two procedures for hypothesis testing, both using the t-test. We will discuss the **One-sample t-test** and the **Independent-samples t-test.**

These procedures owe their names to the t-distribution, a distribution that resembles the normal curve, but that is more appropriate when the sample is small (less then 30 individuals). It is widely used in psychology, where experiments are often conducted on small samples.

The One-Sample t-Test

Review of the method

In the **One-sample t-test**, you want to test the hypothesis that the mean μ of the whole population **differs** from a certain value, which is determined by previous experience, or from a similar situation. For instance, if you know by experience that the average grade in a given course is 77 out of 100 in a group of schools, and you want to test whether a given class differs significantly from that average, you set:

$$H_0: \quad \mu = 77$$
$$H_1: \quad \mu \neq 77$$

If the mean of your sample differs slightly from 77, you do not have a good enough reason to reject the hypothesis, and you can explain the small difference as being due to chance: a random sample is indeed likely to differ slightly from the whole population. But if the difference is large, you conclude that it is not due to chance: if that sample is representative, it must reflect that fact that the population mean is **likely** to be different from 77. But how large is 'large'? Where is the cut-off point beyond which we say: the difference between the observed sample mean and the assumed population mean is too large to be due to chance? Since that whole procedure is based on likelihood, the answer will depend on the risk you are willing to take in drawing your conclusions. Let us say you are willing to take a 5% risk of being wrong. SPSS will tell you how likely it is that the mean of a representative sample will be as far from the assumed population mean as the mean of the sample you have. This is called the level of significance. Let us see how we run the test in practice.

Example

Suppose you want to test the hypothesis that the average age of the American population is 45 years, and verify your hypothesis by using the random sample given in the **GSS93 subset** data file. You set your hypotheses as:

$$H_0: \quad \mu = 45$$
$$H_1: \quad \mu \neq 45$$

You initiate the procedure by choosing
   Analyze
      Compare Means
         One-Sample T Test…

You get the dialog box illustrated in Figure 14.1. You can see in the figure that the variable **Age of Respondent** has already been placed in the **Test Variable** box and the **Test Value** has been set to **45**. Click **OK**.

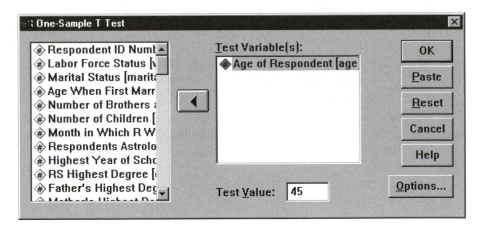

Lab 14 Figure 14.1

   In the Output window, you get Table 14.1.
   The crucial information in Table 14.1 is the column titled **Sig (2-tailed)**, which stands for **Level of Significance**. You interpret it as follows (recall that the table preceding that one in the output gives an average age of 46.23 years and a sample size of 1495 individuals):

• If your population mean is indeed 45 years of age, the probability that you have drawn randomly a sample of 1495 individuals whose mean is equal to 46.23 is 0.007, or 0.7%.

Table 14.1 **One-sample-test**

| | t | df | Test Value = 45 | | | |
| | | | Sig (2-tailed) | Mean Difference | 95% Confidence Interval of the Difference | |
					Lower	Upper
Age of Respondent	2.723	1494	.007	1.23	.34	2.11

- This means that if you assert that the general population mean is 45, it will only happen 0.7% of the time that the sample mean is that far from the assumed population mean (a difference of 1.23 years). This is so rare (less than 1% of the time) that it is safer to say that such a difference indicates rather that your assumption is more likely to be wrong. In doing such a reasoning, you run a risk of 0.7% of being wrong, since this is the probability that you get such a sample when the population mean is 45. You conclude therefore that the hypothesis $\mu = 45$ is put into question by what we found in the sample. The Null Hypothesis is thus rejected, with a 0.007 probability of making a Type I error.

Conclusion: H_0 is rejected since the calculated level of significance is less than 0.05. We concluded that the average age of the population as a whole is not 45 years.

The Independent-samples t-Tests

Review of the method

The **Independent-samples t-test** helps us determine whether two samples, drawn independently, are likely to come from the same population. In other words, we assume that:

> Sample 1 comes from a population with mean μ_1, and that
> Sample 2 comes from a population with mean μ_2.

We then make the hypothesis that $\mu_1 = \mu_2$, or, equivalently, that

$$H_0: \mu_1 - \mu_2 = 0$$

This last statement means that the samples come from populations with identical means; they could even belong to the same population. SPSS will compute the difference of means between the two samples, and will calculate a level of significance. The following example will help us interpret the results.

Example

We want to test the hypothesis that the difference between men and women in our sample on the variables **age** and **rincome91** are significant, that is, they reflect a real difference at the level of the whole population.

1. Select **Independent Samples T Test** (under **Analyze**, then **Compare Means**). You get the dialog box shown in Figure 14.2.

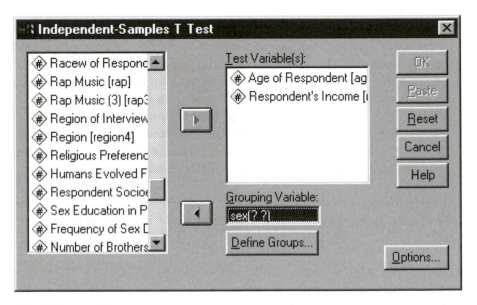

Lab 14 Figure 14.2

2. Place the variables **age** and **rincome91** in the appropriate boxes as shown in Figure 14.2. You should realize that you are running two hypothesis tests at the same time, one for each variable. SPSS allows you to do that.
3. Click on the **Define Groups...** box. You get the dialog box shown in Figure 14.3.
4. In this dialog box, type **1** (for Men), and **2** (for Women). You may have noticed the phrase 'cut point' in this dialog box. T-tests can also be used to compare two groups defined by the fact their members scored above or below a certain value (the cut-off point) on a quantitative variable.
5. Click **Continue**, then **OK**. You get Table 14.2 (some columns of the table have been deleted).

The crucial information here is the calculated level of significance, denoted by **Sig. (2-tailed)**. The results are calculated in two cases: the case where the variances of

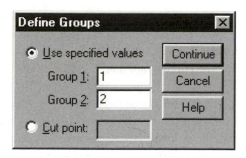

Lab 14 Figure 14.3

Table 14.2 **Independent-samples t-test**

		t	Sig (2-tailed)	Mean Difference	95% Confidence Interval of the Difference	
					Lower	**Upper**
Age of Respondent	Equal variances assumed	−1.697	.090	−1.54	−3.33	.24
	Equal variances not assumed	−1.708	.088	−1.54	−3.32	.23
Respondent's Income	Equal variances assumed	7.470	.000	2.59	1.91	3.28
	Equal variances not assumed	7.488	.000	2.59	1.91	3.27

*(header spanning: **t-Test for Equality of Means**)*

the sub-populations of men and women are identical, and the case where the variances are not identical. SPSS offers tests to determine whether these variances are identical or not, but the discussion of these tests goes beyond the level of this book.

As a practical rule, consider that the variances are equal: Type I errors will be less likely.

In concrete terms, here is how we interpret the results of this table:

* Variable **Age of Respondent**: In this case the null hypothesis is that there is no difference between the ages of men and women in the population at large. We assume that the population of men and women have the same variance for the variable age. If we assert that the difference between men and women is significant, we are taking a 9% risk of being wrong. That is because the difference between their means is very small: 1.54 year. It is too risky to say that such a small difference for that sample reveals a real difference at the level of the whole population. We should rather explain the difference by chance: it is more likely that samples selected independently will display such difference, even if they come from the same population. *Therefore H_0 in this test is accepted.* We

conclude that *we do not have a sufficient reason to reject the hypothesis that men and women in the whole population have the same average age.*

- Variable **Respondent's Income**: This is a different hypothesis test, since the variable is different. The null hypothesis is that there is no difference between the incomes of men and women in the population at large. The incomes are coded as categories, and the mean difference between the average scores of men and women is 2.59. This is a relatively important difference. The SPSS results confirm this understanding: we take a risk that is practically null (rounded up to less than 0.000) when we assert that this difference is significant. Therefore, we can conclude that there is a real difference between the incomes of men and women at the level of the whole population not just for that sample of 1500 people. *Therefore H_0 in this test is rejected and H_1 accepted with a risk less than 0.000.*

Practical rule for interpreting the calculated level of significance:

If the level of significance calculated by SPSS, **Sig.(2-tailed)**, is lower than level of significance we have set (that is, the risk we are willing to take), we reject H_0 and accept H_1. If **Sig.(2-tailed)** is larger than the level of significance we have set (that is, the risk we are willing to take), we accept H_0 and reject H_1.

A final point must be made on the confidence intervals given in the SPSS output. These confidence levels are not confidence intervals for the variable itself, but confidence intervals for the t-statistic that has been calculated. In Table 14.2, we have two situations:

- For the variable **Age of Respondent**, the t-statistic falls *within* the predicted confidence interval. This is why we accept H_0: rejecting it would imply a risk of error of 9%. This is true whether the variances for men and women are equal or not.
- For the variable **Respondent's Income**, the t-statistic falls *outside* the predicted confidence interval. This is why we reject H_0: rejecting it implies a risk of error which is practically null. This is true whether the variances for men and women are equal or not.

Exercises

Pick any **quantitative** variables in any of the data files provided. Run an **Independent-samples t-test** to compare the means of men and women, following the same steps as above. Establish whether the observed differences in the samples reflect a real difference in the whole population, indicating the risk you are taking when you reject the null hypothesis (we have not seen how to determine the risk of being wrong when we accept H_0).

APPENDICES

APPENDIX 1
TABLE OF AREAS UNDER THE NORMAL CURVE

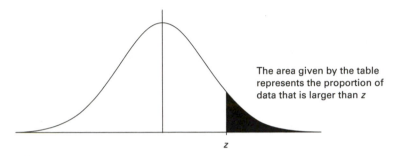

The area given by the table represents the proportion of data that is larger than z

The table gives the proportion of data entries that are larger than the given z-value. For example, 0.2981 of the data is larger than $z = 0.53$

Z	0	1	2	3	4	5	6	7	8	9
0.0	.5000	.4960	.4920	.4880	.4840	.4801	.4761	.4721	.4681	.4641
0.1	.4602	.4562	.4522	.4483	.4443	.4404	.4364	.4325	.4286	.4247
0.2	.4207	.4168	.4129	.4090	.4052	.4013	.3974	.3936	.3897	.3859
0.3	.3821	.3783	.3745	.3707	.3669	.3632	.3594	.3557	.3520	.3483
0.4	.3446	.3409	.3372	.3336	.3300	.3264	.3228	.3192	.3156	.3121
0.5	.3085	.3050	.3015	.2981	.2946	.2912	.2877	.2843	.2810	.2776
0.6	.2743	.2709	.2676	.2643	.2611	.2578	.2546	.2514	.2483	.2451
0.7	.2420	.2389	.2358	.2327	.2296	.2266	.2236	.2206	.2177	.2148
0.8	.2119	.2090	.2061	.2033	.2005	.1977	.1949	.1922	.1894	.1867
0.9	.1841	.1814	.1788	.1762	.1736	.1711	.1685	.1660	.1635	.1611
1.0	.1587	.1562	.1539	.1515	.1492	.1469	.1446	.1423	.1401	.1379
1.1	.1357	.1335	.1314	.1292	.1271	.1251	.1230	.1210	.1190	.1170
1.2	.1151	.1131	.1112	.1093	.1075	.1056	.1038	.1020	.1003	.0985
1.3	.0968	.0951	.0934	.0918	.0901	.0885	.0869	.0853	.0838	.0823
1.4	.0808	.0793	.0778	.0764	.0749	.0735	.0721	.0708	.0694	.0681
1.5	.0668	.0655	.0643	.0630	.0618	.0606	.0594	.0582	.0571	.0559
1.6	.0548	.0537	.0526	.0516	.0505	.0495	.0485	.0475	.0465	.0455
1.7	.0446	.0436	.0427	.0418	.0409	.0401	.0392	.0384	.0375	.0367
1.8	.0359	.0351	.0344	.0336	.0329	.0322	.0314	.0307	.0301	.0294
1.9	.0287	.0281	.0274	.0268	.0262	.0256	.0250	.0244	.0239	.0233
2.0	.0228	.0222	.0217	.0212	.0207	.0202	.0197	.0192	.0188	.0183
2.1	.0179	.0174	.0170	.0166	.0162	.0158	.0154	.0150	.0146	.0143
2.2	.0139	.0136	.0132	.0129	.0125	.0122	.0119	.0116	.0113	.0110
2.3	.0107	.0104	.0102	.0099	.0096	.0094	.0091	.0089	.0087	.0084
2.4	.0082	.0080	.0078	.0075	.0073	.0071	.0069	.0068	.0066	.0064
2.5	.0062	.0060	.0059	.0057	.0055	.0054	.0052	.0051	.0049	.0048
2.6	.0047	.0045	.0044	.0043	.0041	.0040	.0039	.0038	.0037	.0036
2.7	.0035	.0034	.0033	.0032	.0031	.0030	.0029	.0028	.0027	.0026
2.8	.0026	.0025	.0024	.0023	.0023	.0022	.0021	.0021	.0020	.0019
2.9	.0019	.0018	.0018	.0017	.0016	.0016	.0015	.0015	.0014	.0014
3.0	.0013	.0013	.0013	.0012	.0012	.0011	.0011	.0011	.0010	.0010

APPENDIX 2
TABLE OF RANDOM NUMBERS

17031	15532	16006	89840	44231	55053	57859	98136	97770	69560
46068	23960	51491	69217	69235	52827	19263	89194	24726	67945
29287	08655	33171	62907	54727	75276	94979	95023	33087	05835
13452	95111	24993	18476	45482	16698	38247	59071	26588	48372
54822	41406	14108	61813	81646	97975	26103	73433	15984	12248
61157	86506	36521	74559	95909	83124	58410	92789	40796	26484
06297	36767	53523	14324	87207	05735	81259	21006	37041	18175
06380	50714	29468	36676	92934	32241	33596	58725	54268	29531
16922	55536	02285	32303	12394	89836	73410	98591	00932	98470
75522	42808	52924	24688	24178	84663	78611	45748	11730	64919
77676	33246	65833	68819	01858	50522	94887	45814	43572	18939
92087	86889	84083	14650	08095	95590	97963	09335	97474	87457
12320	78625	16733	63782	45620	66638	71798	64974	61135	82106
52356	13378	53650	93061	87111	49382	64647	44638	41368	19047
07901	28539	89636	10385	62725	06591	28382	55094	88209	76728
56343	35970	02447	06844	40971	85887	47863	46252	37497	58570
58144	15741	01028	08871	01297	10732	74618	63693	67541	30104
99547	74293	07796	27114	58187	02955	91212	15016	99462	87670
09414	32790	84727	45600	94054	69968	96711	03620	13178	61944
04277	01667	72142	36426	97145	83220	16067	10268	19617	36639
07364	66305	45505	68845	96583	22892	42200	68326	18363	77033
65558	07846	35533	50124	72186	22898	50040	06093	98906	39580
70782	33400	30616	78710	92467	94177	57873	04405	52316	98513
00265	05446	80843	50570	66276	17289	82648	35778	83999	52254
23675	13661	97400	11151	28082	41771	32730	94989	04214	13908
80817	08512	43918	39234	45514	39559	44774	89216	28606	40950
28426	57962	36149	23071	01274	36321	42625	52913	92319	46602
21699	83058	78365	90291	41255	88742	90959	71911	12364	36500
76196	25126	75243	06455	09699	27259	87118	61898	34070	30663
29910	23090	03597	94026	22920	24888	12060	94269	26857	86418
88434	97038	07584	91689	20272	28419	31588	02629	10304	40897
19631	04682	73840	11159	07635	44424	73856	59622	19881	11576
09207	26807	35382	28345	56942	55028	05253	20412	32181	29668
47677	11154	22639	37745	18850	27436	46074	24186	31239	89847
72402	94986	61941	09619	47521	28566	64866	59213	76003	16418
70120	45782	18412	43100	08253	60693	24399	31778	74211	35853
36947	28111	78904	57756	84812	72046	90147	86847	85179	93120
50512	87468	29926	36505	30657	18856	76880	35595	73896	04326
62112	73064	46212	02202	73464	48659	10954	02926	00794	24981
82353	20212	18866	97787	13678	63246	68120	39120	33821	86064

GLOSSARY OF STATISTICAL TERMS

Acceptance region   The set of all values of the test statistic for which a null hypothesis is accepted.

Alternative hypothesis   The hypothesis which is accepted when the null hypothesis is rejected. It is usually denoted H_A or H_1.

Bell-shaped distribution   A distribution having the overall shape of a bell; normal distributions are among those having this characteristic.

Case   The data in a file that refers to one individual or one unit.

Cell frequency   In a two-way table, it is the number of items falling into an individual cell.

Census   A survey based on the complete enumeration of a population.

Class interval   The range of values covered by a class when coding or grouping a quantitative variable. For any one class, it is given by the difference between its upper and lower boundaries.

Class limit   The upper and lower limits of a class are, respectively, the largest and smallest values it can contain. See also Frequency distribution.

Class mark   The midpoint of a class; hence, the mean of its boundaries or the mean of its limits. See also Frequency distribution.

Codebook   The collection of all the information that allows us to record and interpret the data pertaining to the various variables in a data file. It includes the variable type and format, the variable label, the value labels, and the codes used for missing values together with their meanings.

Confidence interval   The estimate of a parameter, formulated as a whole range of possible values. The probability that the confidence interval will indeed include the parameter is called the *degree of confidence*, or *confidence level*. The formulation of the estimate as an interval is referred to as the *interval estimate*.

Contingency table   A table consisting of two or more rows and two or more columns, also called a *two-way table*. Each row corresponds to a category of the first variable, and each column to a category of the second variable. The cells are the intersection of a row and a column. The numbers written in the cells are the *observed frequencies*, and the numbers at the bottom and side of the table are called the *marginal totals*.

Correlation   The term denotes the relationship (association or dependence) between two or more *quantitative* variables. It is sometimes loosely used to denote statistical association, even when the variables are qualitative.

Correlation coefficient   A measure of the linear relationship between two quantitative variables, known also as the *Pearson product-moment coefficient of correlation*. It is denoted by the letter r and its values range from -1 to $+1$, where 0 indicates the absence of any linear relationship, and -1 and $+1$ indicate, respectively, a perfect *negative* (inverse) and a perfect *positive* (direct) relationship.

Critical region   For a given hypothesis test, the set of values for which the null hypothesis is rejected. Also called the *rejection region*.

Critical values   The dividing values between the acceptance region and the rejection region. Also called *cut-off points*.

CSV format   Comma Separated Variable format. The format of a file of individual values where the values pertaining to different variables are separated by a comma. Another type of format is the ASCII format.

Cumulative distribution   For grouped data, a distribution showing how many of the items are 'equal to or less than' a given value (*ascending* cumulative distribution). It is called a *descending* cumulative distribution if it gives the number of items equal to or larger than the various values.

Data   Data is information which is collected in a systematic way, and organized and recorded in such a way that it can be interpreted correctly. See also Primary data; Raw data; Secondary data.

Data file   A file in paper or electronic format that contains all the data collected in a research or used in a statistical analysis.

Data reduction   The process of summarizing a large amount of data by the methods of descriptive statistics, namely, by grouping them into tables or representing them by means of statistics such as the mean or the standard deviation.

Dependent variable   In a statistical analysis, it is the variable which we try to explain as a function of other variables, called the *independent variables*. One of the objectives of a statistical analysis is to predict the values of the dependent variables in terms of the values of the independent variables.

Descriptive statistics Any treatment of data which aims at summarizing it and presenting it in a way that facilitates its interpretation. Descriptive statistics does not involve generalizations. It includes *measures of central tendency*, *measures of variation*, *measures of position*, *measures of association*, *frequency tables*, and various *graphical representations*, as well as measures of *skewness* and *kurtosis*.

Deviation from the mean The amount by which an individual observation differs from the mean; thus, the deviation of the ith observation x_i from the mean μ_x is given by $x_i - \mu_x$. An important property of deviations from the mean is that, for any set of data, their sum is always equal to zero.

Dichotomy A classification which divides the elements of a population (or sample) into two categories.

Dispersion The extent to which the data values are spread out, or vary from each other. Measures of dispersion could also be called *measures of variation*. The standard deviation is a measure of dispersion.

Distribution A list of the observed values of a variable, together with the frequencies of their occurrence. It could also describe an abstract, theoretical pattern of values such as normal distributions.

Estimate The proposed value for a parameter, computed on the basis of a known sample statistic. It is usually formulated as an *interval estimate*, that is, a whole range of likely values for the parameter, and it is accompanied by a *confidence level*, that is, a probability that the interval estimate does include the parameter. Sometimes, the midpoint of the interval is proposed as a *point estimate* and it is accompanied by a *margin of error*.

Expected (or theoretical) frequency A cell frequency calculated on the basis of some appropriate assumptions in the analysis of the distribution of the data relating to a given variable. For example, for a contingency table, the expected frequency of a cell is the count we would have in that cell if the entries in the various cells were distributed with the same proportions as they are in the marginal totals.

Experimental design The set of planned operations to be performed as part of an experiment.

Frequency The number of items, or cases, falling (or expected to fall) into a category or classification. See also Expected frequency; Observed frequency.

Frequency distribution   The pattern of values obtained when a set of data has been grouped into classes, together with the corresponding frequencies, that is, the number of items falling into each class. Usually presented as a table and called a *frequency table*. It could also refer to the frequencies associated with ungrouped discrete data.

Hypothesis   An assertion about the parameter of a population. For example, the hypothesis that a sample comes from a population with the mean μ, or the hypothesis that a sample comes from a normal population, or that the variances of two populations are equal. See also Alternative hypothesis; Null hypothesis.

Hypothesis testing   A procedure that allows us to test whether a given hypothesis is likely to be true. An assertion about a parameter is proposed, then checked against the results of an empirical observation. If the results are 'too far' from the values proposed in the hypothesis, the hypothesis is rejected, with a certain probability of error. Otherwise, it is accepted (again, with a probability of error).

Independent variable   See Dependent variable.

Kurtosis   A numerical measure of the degree to which a distribution is flat or peaked.

Level of significance   In a significance test, the probability of rejecting a null hypothesis which is true in reality; it is the probability α of committing a Type I error.

Linear regression   See Regression.

Mean   The mean of n numbers is their sum divided by n. Technically, it is called the *arithmetic mean* but the term 'average' is also used in everyday language. The mean is by far the most widely used measure of the central tendency. Population means are usually denoted by the Greek letter μ (mu).

Mean deviation   A measure of the variation of a set of data, also called the *average deviation*, which is defined as the mean of the absolute deviations from the mean.

Median   For ungrouped data, the value of the middle entry (if the number of entries is odd) or the mean of the values of the two middle entries (if the number of entries is even) when the data is arranged in increasing order.

Modal class   The class of a frequency distribution which has the highest frequency; sometimes a class of a frequency distribution which has a higher frequency than adjacent classes.

Mode   The value which occurs with the highest frequency. A distribution can have more than one mode, or no mode at all when no two values are alike.

Negative correlation   Two quantitative variables are said to be negatively correlated when individuals (or units) scoring high on one variable tend to score low on the other variable. In this case the correlation coefficient is negative.

Normal curve   The graph of a normal distribution; it has the shape of a vertical cross-section of a bell. It is symmetric, unimodal, and is determined by a specific mathematical equation. Theoretically, it extends over all real values, but in practice only the values between –3 and 3 are considered.

Normal distribution   A distribution which can be represented by a normal curve. When $\mu = 0$ and $\sigma = 1$, it is referred to as the *standard normal distribution*, or N(0, 1). The proportion of values which lies between two given values can be calculated with the help of a table.

Null hypothesis   A hypothesis H_0 which is to be tested against an alternative hypothesis H_1 (or H_A).

Observed frequency   The actual number of data values falling into a class of a distribution or into a cell of a contingency table.

One-tailed test   A test of a statistical hypothesis in which the region of rejection (the critical region) consists of either the right-hand tail or the left-hand tail (but not both) of the sampling distribution of the test statistic. Correspondingly, a test is referred to as a *two-tailed test* (or a *two-sided test*) when the region of rejection consists of both tails of the sampling distribution of the test statistic.

Open class   A class at the lower or upper end of a frequency distribution having no stated lower or upper limit. For instance, classes such as 'less than $100' or '$75,000 or more' are open classes.

Outliers   Observations at either extreme (small or large) of a distribution which are very far from the main body of the data. Outliers should not be disregarded automatically because they may be very significant in understanding a given distribution.

Parameter   A numerical quantity (such as the mean or standard deviation) which refers to a population. Parameters are usually denoted by a Greek letter to distinguish them from the corresponding description of a sample (called a statistic).

Percentiles   The percentiles $P_1, P_2, ..., P_{99}$ are values that separate the ordered list of values into 100 equal parts.

Pie chart   A pie chart consists of a circle subdivided into sectors whose sizes are proportional to the quantities or percentages they represent.

Point estimation   The estimation of a parameter by assigning it a unique value, called a *point estimate*, together with a possible margin of error.

Population   The set of all units under study. Some authors refer to the set of all units as the *universe*, and reserve the term *population* for the set of all possible values of a given variable.

Population size   The number of elements in a finite population; it is usually denoted by the letter N.

Positive correlation   Two variables are said to be positively correlated, when individuals scoring high on one variable *tend* to score high on the other. In this case, the correlation coefficient is positive.

Primary data   Statistical data which are published by the same organization by which they are collected.

Probability sample   A sample obtained by a method in which every element of a finite population has a known (not necessarily equal) chance of being included in the sample.

Quantitative methods   Quantitative methods are procedures and techniques used to analyze data numerically; they include a study of the valid methods used for collecting data, as well as a discussion of the limits of validity of any given procedure, and of the ways the results are to be interpreted.

Quartiles   The quartiles Q_1, Q_2, and Q_3 are values that separate the set of ordered values of a distribution into four equal parts. The second quartile Q_2 is the same as the median.

Questionnaire   A list of questions given to a subject in a survey.

Quota sampling   A type of judgment sampling, in which the interviewer has to select by any means interviewees with certain characteristics in set proportions; there is no attempt to randomize the selection of subjects within each group.

Random numbers   Published tables of random numbers (or random digits) consist of pages on which the digits 0, 1, 2 … 9 occur with the same probability as if they had been generated completely at random a large number of successive times.

Random sample   A sample chosen so that all the units in the population have the same probability of being included in it; also referred to as *simple random sample*.

Range   The difference between the largest and the smallest values of a set of data.

Raw data   Data which has not been subjected to any sort of statistical treatment (such as grouping, recoding, selecting, etc.).

Regression line   A graphical presentation of the *trend* of a relationship between two variables.

Region of rejection   See Critical region.

Replacement sampling   A method for selecting a sample where each element is replaced before the next one is drawn.

Representative sample   A sample which displays the same characteristics as the population from which it came and that are relevant to a given study, and in similar proportions.

Sample   A part of a population. See Random sample; Representative sample; Sample design.

Sample design   A detailed plan for obtaining a sample from a given population. Also called *sampling plan*.

Sample size   The number of observations (or units) in a sample; it is usually denoted by the letter n.

Sampling   The process of obtaining a sample.

Sampling distribution   The distribution of a statistic, that is, the set of values of the statistic on all possible samples of a given size.

Sampling frame   The list of all units from which the sample is to be chosen.

Scaling   Scaling is a process of measuring, that is, of determining the set of values (scale) that will be used to record information about a variable.

Scatter plot (or scatter diagram)   The set of points obtained by plotting paired measurements as points in a plane. The X-value represents the score of each unit on the independent variable, and the Y-value represents its score on the dependent variable.

Secondary data   Statistical data which is published by an individual or an organization other than the one that collected the data.

Significance level See Level of significance.

Size of a population The number of elements in a finite population; it is usually denoted by the letter N.

Skewness A measure of the lack of symmetry in a distribution. A distribution is said to be *positively skewed* when it contains a larger proportion of data towards its right end than its left end. The coefficient of skewness is then positive. A distribution is said to be *negatively skewed* when it contains a larger proportion of data towards its left end than its right end. The coefficient of skewness is negative in this case.

Spurious correlation This term is used when a correlation between two variables can be attributed to two unrelated processes, that is, there is no causal relationship between the variables.

Standard deviation The standard deviation is given by the square root of the sum of the squared deviations from the mean, divided by the number of units N. It is by far the most widely used measure of the variation of a set of data. It is generally denoted by the letter σ when it refers to a population, and by s when it refers to a sample. In this latter case we divide by $n - 1$ instead of n.

Standard error The standard deviation of the sampling distribution of a statistic.

Standard score A standard score, or *z-score,* is a value in the standard normal distribution N(0, 1). A score in any normal distribution can be converted into a z-score by subtracting from it the mean, then dividing by the standard deviation.

Statistic A quantity (such as a mean or a standard deviation) calculated for a sample.

Statistical inference A form of reasoning that aims at generalizing a measure from sample data to population parameters. Statistical inference usually involves a margin of error and a probability of error.

Statistics (a) The set of methods and techniques employed to collect, present, describe, analyze and interpret data; a scientific discipline. (b) A collection of numerical data, such as the data published by any agency that produces primary data.

Stratified random sampling A method of sampling in which the population is divided into subgroups, or *strata*, and specified portions of the total sample are randomly selected from these *strata*. The principal purpose of this kind of sampling is to guarantee that the various population subgroups are adequately represented in the sample.

Student-t distribution See t-distribution.

Survey design A detailed description of the procedures that will be followed in conducting a survey, including the method used for collecting the data.

Symmetrical distribution A distribution is symmetrical if values that are equally distant from the mean have identical frequencies. The curve representing it is divided into two mirror images around its centre. See also Skewness.

Systematic error A non-random error which introduces a bias into all the observations; such errors are generally the result of a poor design of the observation method.

Systematic sample A sample obtained by selecting every *k*th item on a list. The first unit to be chosen in the list is generally chosen at random among the first *k* units.

t-Distribution A distribution which is similar to the normal distribution, and which is used for inference when the variance of the population is unknown or when the sample size is small (less than 30 units).

t-Tests Hypothesis tests based on the t-distribution. The *one-sample t-test* is a hypothesis test to decide if a random sample is likely to come from a normal population with mean $\mu = \mu_0$; it is used when the population standard deviation is unknown. The *two-sample t-test* is a test concerning the difference between the means of two normal populations.

Test of significance A statistical procedure used for deciding whether an observed difference in one or several samples is due to chance, or whether it is significant, that is, it is likely to reflect a *real* difference at the level of the whole population.

Test statistic The statistic on which the decision to accept or reject a given hypothesis is based.

Test of hypotheses (or hypothesis testing) A procedure for deciding whether to accept or reject a hypothesis.

Treatment A treatment is a condition whose effects are to be studied in an experiment. It is applied to an experimental group but not to the control group.

Treatment effect In an experiment, a quantity which represents the change in response produced by a given treatment.

Type I error In hypothesis testing, the error resulting from the rejection of a null hypothesis when it is true in reality. The probability of committing a Type I error is denoted by α.

Type II error   In hypothesis testing, the error resulting from accepting the null hypothesis when it is false in reality. The probability of making a Type II error is denoted by β.

Unimodal   A distribution is said to be unimodal if it has only one mode or modal class.

Validity   The extent to which a test measures what it claims to measure.

Variable   A characteristic that is observed, and recorded in a single column in a data file.

Variable label   The name of the variable that appears when tables are printed out by a statistical software.

Value label   The name used to designate a value of a categorical variable.

Variance   The square of the standard deviation.

Variation   The extent to which observations or distributions are spread out, or dispersed.

Variation, coefficient of   A measure of *relative variation* which is obtained by dividing the standard deviation of a distribution by its mean. It expresses the magnitude of the variation compared to the average size. When it is multiplied by 100, it expresses the relative variation as a percentage.

Variation, measures of   Statistical measures such as the standard deviation, the mean deviation, or the range, which are indicative of how spread out a distribution is.

Weighted mean   The average of a set of numbers obtained by multiplying each number by a weight expressing its relative importance. The sum of weights should add up to 1.

z-Scores   A *z-score*, or standard score, is a value in the standard normal distribution $N(0, 1)$. A score in any normal distribution can be converted into a z-score by subtracting from it the mean, then dividing by the standard deviation.

GENERAL BIBLIOGRAPHY

The references below have been divided into five categories: general references, references on SPSS, references with a disciplinary focus, references on qualitative research, and critical discussions of methodological issues. These references were not included in the suggested reading provided with the various chapters.

General

Black, Thomas R. (1999) *Doing Quantitative Research in the Social Sciences: an integrated approach to research design, measurement and statistics*. London: Sage.

Black, Thomas R. (1993) *Evaluating Social Science Research: An Introduction*. Thousand Oaks, CA: Sage.

Bouma, Gary D. and Atkinson, G.B.J. (1995) *A Handbook of Social Science Research*. New York: Oxford University Press.

Carmines, Edward G. and Zeller, Richard A. (1979) *Reliability and Validity Assessment*. Beverly Hills, CA: Sage Publications.

Bickman, L. and Rog, Debra (eds) (1998) *Handbook of Applied Social Research Methods*. Thousand Oaks, CA: Sage.

Howe, Renate and Lewis, Ros (1993) *A Student Guide to Research in Social Science*. New York: Cambridge University Press.

Hult, Christine A. (1996) *Researching and Writing in the Social Sciences*. Boston: Allyn and Bacon.

Kaplan, Abraham (1964) *The Conduct of Inquiry; Methodology for Behavioral Science*. San Francisco: Chandler Pub. Co.

Katzer, Jeffrey (1998) *Evaluating Information: a Guide for Users of Social Science Research* (4th edn). Boston: McGraw-Hill.

Neuman, William Lawrence (2000) *Social Research Methods: Qualitative and Quantitative Approaches*. Boston: Allyn and Bacon.

Newman, Isadore and Benz, Carolyn R. (1998) *Qualitative-Quantitative Research Methodology: Exploring the Interactive Continuum*. Carbondale: Southern Illinois University Press.

Sproull, Natalie (1995) *Handbook of Research Methods: a Guide for Practioners and Students in the Social Sciences* (2nd edn). Metuchen, NJ: Scarecrow.

Reference Works on SPSS

Carver, Robert H. and Nash, Jane Gradwhol (2000) *Doing Data Analysis with SPSS 10.0*. Pacific Grove, CA: Duxbury.

Green, S.B., Salkind, N.J. and Akey, T.M. (2000) *Using SPSS for Windows, Analyzing and Understanding Data* (2nd edn). Upper Saddle River, NJ: Prentice Hall, 2000.

Norusis, Marija J. (1998) *SPSS 8.0 Guide to Data Analysis*. Upper Saddle River, NJ: Prentice Hall.

General Works With a Disciplinary Focus

Clark-Carter, David (1997) *Doing Quantitative Psychological Research: From Design To Report*. Hove, East Sussex: Psychology Press.

Coolican, Hugh (1990) *Research Methods and Statistics in Psychology*. London: Hodder & Stoughton.

Dale, Angela and Davies, Richard B. (ed.) (1994) *Analyzing Social and Political Change, A casebook of Methods*. London: SAGE Publications.

Harnett, Donald H. and Murphy, James L. (1993) *Statistical Analysis for Business and Economics*. Don Mills, Ontario: Addison-Wesley Publishers.

Mertens, Donna M. (1998) *Research Methods in Education and Psychology: Integrating Diversity with Quantitative & Qualitative Approaches*, Thousand Oaks, CA: Sage Publications.

Michell, Joel (1999) *Measurement in Psychology: Critical History of a Methodological Concept*. Cambridge, UK; New York, NY: Cambridge University Press.

Reis, Harry T. and Judd, Charles M. (ed.) (2000) *Handbook of Research Methods in Social and Personality Psychology*. New York: Cambridge University Press.

Sommer, Barbara (1999) *A Practical Guide to Behavioral Research: Tools and Techniques* (4th edn). New York: Oxford University Press.

Thyer, Bruce A. (ed.) (2001) *The Handbook of Social Work Research Methods*. Thousand Oaks, CA: Sage Publications.

Qualitative Research

Berg, Bruce Lawrence (1998) *Qualitative Research Methods for the Social Sciences* (3rd edn). Boston: Allyn and Bacon.

Coffey, Amanda (1996) *Making Sense of Qualitative Data: Complementary Research Strategies*. Thousand Oaks, CA: Sage Publications.

Denzin, Norman K. and Lincoln, Yvonna S. (eds) (2000) *Handbook of Qualitative Research* (2nd edn). Thousand Oaks, CA: Sage.

Marshall, Catherine and Rossman, Gretchen B. (1999) *Designing Qualitative Research*. Thousand Oaks, CA: Sage Publications.

Maykut, Pamela S. and Richard Morehouse (1994) *Beginning Qualitative Research: A Philosophic and Practical Guide*. Washington, DC: Falmer.

Miles, Matthew B. and Huberman, A. Michael (1994) *Qualitative Data Analysis: An Expanded Sourcebook*. Thousand Oaks, CA: Sage.

Critical Discussions

Maier, Mark H. (1995) *The Data Game, Controversies in Social Science Statistics* (2nd edn). New York: M.E. Sharpe.

Reinharz, Shulamit (1992) *Feminist Methods in Social Research*. New York: Oxford University Press.

Renzetti, Claire M. and Raymond M. Lee (eds) (1993) *Researching Sensitive Topics*. Newbury Park, CA: Sage.

Seltzer, Richard A. (1996) *Mistakes That Social Scientists Make: Error and Redemption in the Research Process*. New York: St. Martin's.

Williams, Frederick (1992) *Reasoning with Statistics: How to Read Quantitative Research*. Fort Worth: Harcourt Brace Jovanovich College Publishers.

INDEX